DIMITRIS SALAPATAS

THE AEGEAN SEA DISPUTE BETWEEN GREECE AND TURKEY

The consequences for NATO and the EU

Ακακία

Published in England by AKAKIA Publications, 2014

Dimitris Salapatas

THE AEGEAN SEA DISPUTE BETWEEN GREECE AND TURKEY

The consequences for NATO and the EU

ISBN: 978-1-909884-52-6

Cover Image:
Source: ShutterStock.com / Contributor: Suns07Butterfly / File No: 166222079
Mixed and Designed by AKAKIA Publications

Akakia
PUBLICATIONS

St Peters Vicarage, Wightman Road, London N8 0LY, UK

T. 0044 203 28 66 550
T. 0044 203 28 96 550
F. 0044 203 43 25 030
M. 0044 7411 40 6562

www.akakia.net
publications@akakia.net

2014, London, UK

CONTENTS

Στους γονείς μου
To my parents

LIST OF ABBREVIATIONS

EEC	European Economic Community
EEZ	Exclusive Economic Zone
EU	European Union
FIR	Flight Information Region
G20	Group of 20
ICAO	International Civil Aviation Organisation
LOS	Law of the Sea
NAC	North Atlantic Council
NATO	North Atlantic Treaty Organisation
NOTAM	Notice to Airmen
UN	United Nations

Istanbul
TURKEY
GREECE
Aegean Sea
Athens

ABSTRACT

The Aegean dispute between Greece and Turkey is a persistent problem between the two allied states. Difference of interpretation of the treaties has contributed in the prevalence of the argument. This dispute consists of five key issues. Greece only accepts one, namely the delimitation of the continental shelf. However, Turkey has introduced and has persisted on the other four, which are the delimitation of the territorial seas, the national airspace and FIR controls over the Aegean Sea, the demilitarization of the Eastern Aegean Islands and finally the disputed islands, islets and rocks which have presented the grey zones issue. All of these matters have persisted for so long, especially after the Treaty of Lausanne in 1923 where the current status quo in the Aegean was introduced, due to economic, political and strategic reasons.

The Aegean dispute does not only have consequences for Greece and Turkey; it also affects, negatively, NATO and the European Union. This is an ongoing problem, which if not solved it will produce future problems, not only for the two states, but also for NATO and the EU. A third party may be needed in order to assist in finding a just and permanent solution concerning this dispute, since Greece and Turkey seem incapable of solving this dispute by themselves.

This book is based on a thesis, which was submitted to the University of London, Queen Mary, for the MA in International Relations, 2010. Since then the Aegean Sea dispute has remained unchanged for both Greece and Turkey. However, the current form of the text has altered, due to a number of new information. Unfortunately, a solution to this significant matter has yet to be found.

ACKNOWLEDGEMENTS

I would like to thank my family and friends for their immense support during my academic studies, for the production of this book and their continuous assistance in every aspect of my life. Without their encouragement I would not be the person I am today and this book would not exist. I would like to thank my supervisor for my MA in International Relations, Dr Patricia Owens (Senior Lecturer in the School of Politics and International Relations, Queen Mary, University of London) who assisted me in forming the final title and structure of this paper. I am very grateful to H.E. the Ambassador of the Hellenic Republic, currently to Astana – Kazakhstan - , Mr Makis Pantzopoulos, who helped me greatly in understanding the delicate issues and disputes between Greece and Turkey. Also, I would like to express thanks to Mr Savvas Kalenteridis (former Greek Intelligence Officer and presently Publisher), Angelos Syrigos (Assistant Professor of International Law & Foreign Policy, Department of European International and Area Studies, Panteion University, Athens) and Mr Makis Galiatsatos (European Parliament) for assisting me, by pointing me towards the right direction, in regards to crucial sources for the completion of this academic work. Finally, I would like to thank Akakia Publications for accepting to publish this book.

INTRODUCTION
A Brief Summary of the Dispute

"Few boundary regions have been fought over more arduously or for so long as the one separating Greece and Turkey"[1]. The two states have a unique relationship within the international arena. There relations have included conflicts and opposite interests, which have made the South-Eastern part of Europe an unstable region. Although being allies in NATO, Greece and Turkey have not found a way to abolish their differences, whilst a change of the status quo could drive the two states to war.

The Aegean Sea, one of the key disputed topics between Greece and Turkey, is an important issue, as Minister of Foreign Affairs of Turkey, Ahmet Davutoglu explains, "The Aegean is the major marine focal point of the Eurasia universal continent in the direction North-South. In contrast to the cases of the Red Sea and the Persian Gulf, which occupy similar positions, the Aegean, which is located in close proximity from the three continents and provides the possibility of open passage towards the three (continents) without a land obstruction, it has from this perspective a first class strategic importance not

[1] International Boundary Study, "Greece-Turkey Boundary", http://www.law.fsu.edu/library/collection/LimitsinSeas/IBS041.pdf, accessed 28-07-2010, 19.02, p. 2

only for Greece and Turkey, which have a coast line in this sea, but also first of all for the countries in the Black Sea and also for the global and peripheral powers which have a need of a transport and commercial hub"[2].

"The fact that these two states earned their national identities by fighting each other has undoubtedly affected their subsequent relations"[3]. The Aegean dispute is a chronic problem for both countries. Each state understands this dispute differently. Greece only accepts the "delimitation of the Continental Shelf, which is the only legal

---

[2] «Το Αιγαίο αποτελεί το σημαντικότερο θαλάσσιο κομβικό σημείο της Ευρασιατικής παγκόσμιας ηπείρου στην κατεύθυνση Βορρά-Νότου. Σε αντίθεση με την περίπτωση της Ερυθράς θάλασσας και του Περσικού κόλπου, που κατέχουν παρόμοιες θέσεις, το Αιγαίο, που βρίσκεται σε μία κοντινή απόσταση και από τις τρεις ηπείρους και παρέχει τη δυνατότητα του ανοίγματος και στις τρεις χωρίς κάποιο χερσαίο εμπόδιο, έχει από αυτήν την άποψη μία πρώτης τάξης στρατηγική σημασία όχι μόνο για την Ελλάδα και την Τουρκία, που διαθέτουν ακτές σε αυτή τη θάλασσα, αλλά κατ'αρχάς για τις παρευξείνιες χώρες και έπειτα για όλες τις παγκόσμιες και περιφερειακές δυνάμεις που έχουν ανάγκη ενός μεταφορικού και εμπορικού κόμβου».
Davutoglu, Ahmet, *To Stratigiko Vathos, I Diethni Thesis tis Tourkias, (Athens, Poiotita, 2010), p.267*

[3] Aydin, Mustafa, "Contemporary Turkish-Greek Relations: Constraints and Opportunities", in Aydin, Mustafa and Kostas, Ifantis (eds), *Turkish-Greek Relations, The Security Dilemma in the Aegean*, (London and New York, Routledge, 2004), p. 23

difference between the two countries"[4], as the only point in the dispute. Turkey on the other hand challenges Greece's right to extend its territorial waters from the current six miles to up to twelve miles. It also contests Greek airspace, the Athens FIR, the militarization of the islands in the Eastern Aegean and promotes the theory of the grey zones by disputing the sovereignty of several Greek islands, which is a direct questioning of the sovereignty rights of Greece in the Aegean.

The consequences for NATO and the EU is a key matter due to the fact that the two states are members of NATO, whilst Greece is a EU member state and Turkey has been trying to enter the Union for quite some time. These two international bodies have had to face some difficult circumstances, due to the existence and prevalence of this dispute which presents problems for both NATO and the EU.

It is vital to identify why this question is important not only for the bilateral relations between Greece and Turkey but also why a scholar and the discipline of International Relations should acknowledge the importance of this dispute. It is interesting to identify how two allies, two European countries, which are members of NATO, are in dispute over so many issues, not only in respect to the Aegean Sea but also on minority issues and the

[4] Ministry of Foreign Affairs, Greece, http://www.ypex.gov.gr /www.mfa.gr/en-US/Policy/Geographic+Regions/South-Eastern+Europe/Turkey/Relationships+-+Potentials/, accessed 26-07-2010, 12.02.

Cyprus problem, which add additional difficulties to the already unstable Greek-Turkish relations. All of these predicaments have brought the two neighbouring states close to the brink of war a number of times in the last 30 years. The Aegean Sea dispute is not only a bilateral matter but an international one too, where NATO, the EU and the USA have had to face many dilemmas and in many cases have intervened to prevent war, as was the case in January 1996 over the Imia crisis.

In the following chapters the diverse views from both Greece and Turkey will be stated to make it clear why this ongoing problem persists. The Treaties, which are the basis for the status quo that exists up to this day in the Aegean Sea, will be analysed to clarify and verify what the two countries support and believe. The Treaty of Lausanne is the key text used by both states to confirm their policies and theories, producing many problems in the process.

For this paper, the main objective is to state the two views and show why this is a persisting dispute between two allied states. The consequences for both NATO and the EU emphasize the international character of this quarrel. In the bibliography used for this book it was not clearly evident how this dispute has consequences for NATO and the EU, making it a first attempt to achieve this goal. I thought it useful and objective to study and present in this paper observations and opinions from both sides of the Aegean. Primary sources have been used, such as NATO and EU, documents from the two

Foreign Ministries as well as from the Hellenic National Defence General Staff which emphasize the difference in approach and understanding of the dispute. Also secondary sources from scholars, politicians, diplomats from Greece, Turkey and other European and NATO states, will illustrate emphatically the true problems and the reasons why this is a chronic thorn in the relations of the two South-Eastern European states. To make this paper more comprehensible, examples and case studies will be given with the aim of demonstrating, in an iconic way, the disputed issues and the consequences they have for Greece, Turkey, NATO and the EU.

In the first part the Aegean dispute will be examined. The five main points are the continental shelf, the width of the territorial seas, the airspace issue and the Athens FIR problem, the demilitarization and militarization of the Eastern Aegean Islands and the disputed islands, islets and rocks, eventually introducing the theory of grey zones. The reasons why this dispute prevails will be evaluated to confirm that it is not a simple topic which could easily be resolved.

The consequences for NATO are stressed on the strategic problems which occur, due to the existence and the pervasiveness of this dispute within the South-Eastern flank of the alliance; furthermore, they also affect the politics and economic aspect of the alliance. Both Greece and Turkey have been members of NATO since 1952 but their constant disputed issues have negative outcomes for the alliance. NATO's South-Eastern flank is vulnerable and is constantly compromised by aggressive military actions, making it a week spot for the security of the alliance. Also taking decisions, especially when a unanimous vote is needed, seems a problematic endeavour. This dispute between two of NATO's members has also highlighted the alliance's neutral attitude towards any argument between Greece and Turkey.

The concerns for the EU are also imperative for the Union, which recognizes the dispute as a regional one which affects many policies within South-East Europe. The consequences are political, economic and strategic. The resolution of this dispute will compensate Turkey, which has wanted to become an EU member for the past decades. The negative consequence of this dispute for Europe is the immigration problem, which affects not only the Union but also each member state individually. Due to the dispute, FRONTEX finds it hard and many times impossible to complete its missions of protecting the EU by policing the Aegean for illegal immigrants. Disputes over grey zones and islands have made it hard for FRONTEX to achieve its goals. Regional stability, which also relates to the border issue, is the second consequence for the EU. South-Eastern Europe is not only a liability for NATO but also for the European Union, when referring to regional safety and stability. Finally the EU's role in this disagreement and its neutral treatment of both Greece and Turkey, concerning the Aegean dispute, is emphatically stated through examples from the last part of the 20th century.

Concluding, it will be evident that a resolution of this chronic dispute between Greece and Turkey over the Aegean is vital, for economic, political and strategic reasons. It will also be a positive move for both states but also for the international organizations they are part of, i.e. NATO and the EU; the negative consequences will be diminished, which will consequently mean that both states will be part of a more peaceful, prosperous and profitable future.

CHAPTER 1
Explaining the Dispute

Introduction

"The current dispute really has its origins in the Treaty of Lausanne"[5]. The basis for the current situation is set in this Treaty, which has also been used in the treaties since then to solve the Aegean Dispute. This issue is fundamental due to the fact that it affects "directly on vital national interests concerning territorial sovereignty and security for both countries"[6]. Each state accepts, understands and interprets the dispute in their own way. Greece only acknowledges the delimitation of the continental shelf as the only problem between the two states. Greece "claims that it should be decided at the International Court of Justice and according to

---

[5] Mann, Stephen, *The Greek-Turkish Dispute in the Aegean Sea: Its Ramification for NATO and the Prospects for Resolution*, (Monterey California, Naval Postgraduate School, 2001), p.3

[6] Aydin, Mustafa, "Cacophony in the Aegean; Contemporary Turkish Greek Relations, University of Ankara", http://www.politics.ankara.edu.tr/eski/dosyalar/MMTY/27/6_mustafa_aydin.pdf, accessed 28/05/2010, 17.46, p. 115

the existing international law"[7], because it recognizes the dispute as being a legal difference. Turkey, on the other hand, puts on the table other concerns which make the resolution of this dispute even more intricate. The latter country wishes to solve this issue through bilateral talks and not through international bodies, as Greece suggests. The five thorny issues will be analysed below, in order to understand the whole dispute.

A. Continental Shelf

The continental shelf is the key issue, from Greece's point of view, that prevails within this disagreement. Before analysing this disputed matter it is important to state the definition of the continental shelf, which is:

"Under Article 76(1) of the LOS Convention, the continental shelf extends beyond its territorial sea to the outer edge of the continental margin or to a distance of 200 nautical miles from the territorial sea baselines, where the continental margin does not extend up to that distance. As a result the geological and geomorphological characteristics of the seabed are not taken into consideration within the 200 miles limit. Contemporary international law does not accept geological or geomorphological characters as

[7] Anastasakis, Othon, "Greek - Turkish relations: Old problems, new challenges, Barcelona Centre for International Affairs" , www.cidob.org, accessed 25/05/2010, 17.35, p. 71

legal basis for title to continental shelf within 200 miles from the coast".[8]

This dispute began in 1973 when for the first time Greece attempted to find in the Aegean seabed oil, minerals and gas. The same year Turkey approved permits to the Turkish Petroleum Company to explore the Aegean Sea for resources. It also "published a map indicating the limits of Turkey's continental shelf rights as being to the west of Greece's easternmost islands"[9]. The dispute over the continental shelf intensified when Greece revealed that it had found oil and natural gas near the island of Thassos, which according to the magazine *Time* the Thassos oil deposit is "of low quality and hardly worth exploiting"[10] . From this period and till 1976 Turkey was constantly sending research ships to explore the seabed in areas maintained by Greece. As a result of this, the two countries came at that time close to war.

---

[8] Strati, Athanasia, "Greece and the Law of the Sea: A Greek Perspective", in Chircop Aldo, Andre Gerolymatos and John Iatrides (eds), *The Aegean Sea After the Cold War* (London, Macmillan Press LTD., 2000), p. 94

[9] Pratt ,Martin and Clive, Schofield, 'The Imia/Kardak Rocks Dispute in the Aegean Sea', *IBRU Boundary and Security Bulletin Spring*, (1996), p. 63

[10] Time Magazine, "THE MEDITERRANEAN: The West's Ragged Edge", 12 June 1978, http://www.time.com/time/magazine /article/0,9171.948171,00.html, accessed 25-05-2010, 18.46

In 1987 the two neighbouring states came close to war once again. Turkey sent Sismik-I into the Aegean, together with Turkish warships to explore the seabed, after Greece showed interest to drill near Thassos, 10 nautical miles from the island, which means that Greece wished to drill in international waters, but within its continental shelf. This of course is not a matter of the past. The Cesme, a Turkish research ship, was in recent years exploring once again the Aegean seabed. It started on 19-07-2010 within the Greek continental shelf, between Thrace and the island of Samothrace. The then Greek Alternate Foreign Minister Mr. Dimitris Droutsas commenting on another Turkish ship, the Piri Reis, which was in the recent years found cruising South of Kastelorizo, said that "...Turkey seeks to create conditions of tension and it exacerbates the situation in the Aegean, pondering the known positions of moving outside international legitimacy and challenging Greek sovereign rights"[11]. Therefore, it is identifiable that this dispute is an on-going and current problem, continuing into the 21st century.

---

[11] "Εδώ και μερικές εβδομάδες η Τουρκία επιχειρεί να δημιουργήσει συνθήκες έντασης και να οξύνει την κατάσταση στο Αιγαίο, αναμασώντας τις γνωστές της θέσεις που κινούνται εκτός διεθνούς νομιμότητας και αμφισβητώντας κυριαρχικά δικαιώματα της Ελλάδας". Kathimerini newspaper, "Diavima apo YPEKS gia 'Tsesme' kai 'Piri Reis', http://www.kathimerini.gr/4dcgi/_w_articles_kathremote_1_20/07/2010_347532, accessed 20-07-2010, 13.47

In January 2013 Greece was planning unilateral drilling in disputed areas, which prompted numerous cautious warnings from Turkey. Additionally, in February 2013 Antonis Samaras (the Greek Prime Minister) claimed that Greece has the right to search for oil and gas, threatening to declare an Exclusive Economic Zone (EEZ)[12], something in which Turkey is against, since this will limit Turkey's territorial waters. Both countries turned to the United Nations, giving their own description of their rights within the region. This is another example of how "Greece is evidently laying its cards on the table and trying to force some movement on the Aegean border issue. Turkey, however, is unlikely to compromise"[13]. Also, Greece wishes to find oil now, due to the fact that it is experiencing an economic crisis (since 2008) and therefore needs to find solutions to its current economic situation.

The dispute over the continental shelf prevails since both countries understand the seabed issue

---

[12] The Exclusive Economic Zone has yet to be delineated by Greece. Turkey, on the other hand, does not accept the EEZ, mainly due to the fact that it will have no entitlements in the Aegean Sea; and therefore will have no rights to any oil or gas found in the Aegean Sea. A key matter to this dispute is Kastelorizo, at the South of Turkey, due to its proximity to Asia Minor. This is most likely the reason why Greece has yet to delineate its EEZ with Cyprus, Egypt and Libya.

[13]Jackson, Alex, "Natural, Gas Europe" http://www.naturalgaseurope.com/greece-and-turkey-spar-over-aegean-energy, accessed 18-09.2013, 20.29

from a different perspective. Greece states that the islands do have a continental shelf. It bases this on Article 121(3) of the LOS Convention where "all islands can generate in principle all maritime zones, that is, the territorial sea, contiguous zone, EEZ and continental shelf"[14]. Greece emphasizes the importance of international law to solve this dispute. Turkey on the other hand understands this clash of interests as being a political one, and so believes that bilateral talks will be the best solution. Turkey does not recognize that the islands obtain a continental shelf, believing that the seabed and the islands are a prolongation of the Anatolian peninsula. This of course makes it impossible for the two countries to delimitate the continental shelf. Nonetheless, all of the difficulties which seem impossible to solve should not prevent the two Aegean states of attempting to solve the dispute. Greece should acknowledge that Turkey has rights in the Aegean and Turkey should not feel threatened by Greece, thinking that the latter is trying to reaffirm the theory that the Aegean is a 'Greek Lake'. As Athanasopulos states, "both nations must address and understand each other's justifiable threats, fears and national security concerns regarding their

[14] Strati, Athanasia, "Greece and the Law of the Sea: A Greek Perspective", in Chircop Aldo, Andre Gerolymatos and John Iatrides (eds), *The Aegean Sea After the Cold War* (London, Macmillan Press LTD., 2000), p. 95

relationship in the Aegean"[15]. This mistrust in bilateral relations is the key aspect to why the two states are in dispute over the Aegean.

B. Territorial Seas

It is crucial to recognise that all the issues within this quarrel are related and most of them are the continuation of another disputed matter. Both Greece and Turkey claim six miles of territorial waters in the Aegean. Greece also obtains the six miles in the Ionian Sea and Cretan Sea, South of Crete. Turkey, on the contrary, obtains a 12 mile territorial claim in the Black Sea and Mediterranean Sea. The reason why Turkey has not extended its territorial waters to 12 miles in the Aegean is because, as Turkish Ambassador Deniz Bolukbasi claims, they are concerned that Greece will immediately do the same.

Greece retains the right to extend its territorial waters to 12 nautical miles according to the "1982 UN Convention on the Law of the Sea which includes, under Article 3, the stipulation that 'Every state has the right to establish the breadth of its territorial sea up to a limit not exceeding 12 nautical miles"[16].

[15] Athanasopoulos, Haralambos, *Greece, Turkey and the Aegean Sea*, (Jefferson, North Carolina, McFarland and Company Inc. Publishers, 2000), p. 55

[16] Pratt ,Martin and Clive, Schofield, 'The Imia/Kardak Rocks Dispute in the Aegean Sea', *IBRU Boundary and Security Bulletin Spring*, (1996), p. 62

Greece has ratified this convention, but Turkey has not. Greece has yet to pragmatize this right. A reason for this is that "Turkey, since 1995 with a unanimous resolution of the Turkish National Assembly, threatens with war in case Greece extends her territorial sea (casus belli)"[17]. This threat towards Greece is an infringement of Article 2(4), of the UN Charter which disallows the use of any force between member states, like Greece and Turkey. However, it should be understood that Turkey has a reason for being hostile towards Greece on this subject. If Greece realizes its right and extends its territorial waters to 12 nautical miles then Turkey will have no access to any high seas from its territory or from the Black Sea. Even so, Greece has confirmed that the right of innocent passage will be enforced if it was to extend its territorial waters. This last statement is not satisfactory to Turkey, which will see its rights in the Aegean being limited. Then the Aegean Sea will become what Turkey has feared, a 'Greek Lake'. On the other hand, most importantly, Greece does not have any support from its allies to extend its territorial waters to 12 nautical miles. The USA is against it, understanding that this increase would interfere with the free passage of its ships and aeroplanes in the region.

[17] Hellenic National Defence General Staff, http://www.geetha.mil.gr/index.asp?a_id=2781, accessed 26-07-2010, 11.56

The extension to 12 nautical miles will, in Turkey's view, "eliminate the Turkish claims and it will positively augment the National Defence of Greece"[18]. Greece will claim in this case 64 per cent of the Aegean, in contrast to the 35 per cent it currently obtains, and Turkey will gain 1 per cent taking it to 10 per cent, in comparison to the 9 per cent it presently covers. This will be a massive advantage for Greece and a huge disappointment for Turkey. Greece would then possess most if not all the oil and natural gases of the Aegean, boosting Greek economy. However, it is imperative to understand that Turkey also has rights in the Aegean Sea; economic, political and national interests which could not be dismissed. Therefore, talks should continue in order to find a solution for not only the two countries, but also for all the allies who pass from the Aegean Sea.

Although the issue of territorial seas seems a straight forward matter, the continental shelf complexes the Aegean dispute; especially when minerals are discovered. Nonetheless, a solution could be found and the extension of territorial sea could be put up for discussion; on the other hand, Turkey cannot threaten a neighbour with war if it was to extend its breadth of territorial sea. The casus belli threat is the key reason why this issue will not

---

[18] Syrmos, Theodoros, *Territorial Waters of Greece and Air/Maritime Navigation*, (Alabama, Maxwell Air Force Base, 2002), p. vi

be resolved in the near future. Both states need to try and find a solution, so they can both take advantage of their minerals within the boundaries which will be delimitated. The first dispute to be solved should be the territorial sea extension. "Having solved this issue, the two countries will have a rigid base on which they can resolve the common accepted dispute of the delimitation of the continental shelf"[19].

C. Airspace

The airspace disputed matter is a crucial one, especially in the modern globalised world where air transportation is part of everyday routine. Two sub-categories are attached to the airspace dispute; the airspace zone, meaning the width of the airspace, and the Flight Information Region (FIR) over the Aegean, which is contested on a daily basis by Ankara.

The airspace is the first disputed matter. Greece since 1931 has established, "with the Presidential Decree of 6/18 September 1931"[20] airspace of 10 nautical miles. After ascertaining it no

[19] Patsiaouras, Konstantinos, *Democratic Peace theory and Greek-Turkish Relations in the Context of the European Union*, (California, Thesis Naval Postgraduate School, 2009), p. 12

[20] Hellenic National Defence General Staff, http://www.geetha.mil.gr/index.asp?a_id=2781, accessed 26-07-2010, 11.56

country, including Turkey, objected in any way. After 1974, and the invasion of Cyprus by the Turkish army, many issues in the Aegean intensified. One of them was the Airspace. Turkey disputes the 10 mile national airspace. That is why, as Mustafa Aydin states, "Turkey frequently exercises 'its right' to fly over international airspace of the Aegean up to a distance of 6 miles to the Greek mainland"[21]. However, what is not stated by Turkey is that it does not only violate Greek airspace up to the 6 miles point, but also sends its military aircrafts to fly over Greek islands. The US Arms Clients Profile emphasises this by showing that in "1996 Turkish jet-fighters violated Greek air space 1.600 times, of which 538 were over-land (mostly islands of the Aegean)"[22]. This prompts Greek complaints that Turkey violates Greek airspace. In the year 2010, up to June, statistics showed that there had been 635 violations of the Greek national airspace. Turkey supports the view that it flies within international airspace and over grey zones, where the islands and rocks are disputed (this will be analysed further down). Even so, it does not explain why Turkish military aircrafts fly over populated islands, such as

---

[21] Aydin, Mustafa, "Contemporary Turkish-Greek Relations: Constraints and Opportunities", in Aydin, Mustafa and Kostas, Ifantis (eds), *Turkish-Greek Relations, The Security Dilemma in the Aegean*, (London and New York, Routledge, 2004), p. 29

[22] U.S Arms Clients Profile-Greece, http://fas.org /asmp/profiles/greece.htm, accessed 14/07/2010, 14.53

Farmakonisi and Agathonisi which are situated in the Dodecanese Region, and also over non-grey zone areas. They wish to make a profound presence in the region. That is why they also fly low, in order to threaten not only the Greek military, but also the citizens of Greece. In September 2013, Kathimerini claimed that six Turkish jets entered the Greek air space, between Samos and Chios. "The jets, four F-4 Phantoms and two F-16s, entered the Athens Information Region (FIR) at 11.28 a.m. and flew over the islet of Anthropophagi, which is part of the Fournoi complex, at 4,000 feet"[23]. This is undoubtedly a continuation and intensification of the threat by Turkey towards Greece and its people.

The countless violations of Greek airspace, over the Aegean Sea and islands, result in daily dogfights between the Greek and the Turkish Air forces. This, inevitably, increases the danger of a collision and an increase of tension between the two states. One of the last crashes between two military aircrafts was on the 23rd of May 2006. One Greek and one Turkish F-16 fighter jets crashed into the Southern Aegean Sea, 21 nautical miles from the Greek island of Karpathos, after colliding in mid-air. This unfortunate event, whereby the Greek pilot (Captain Konstantinos Iliakis) was killed in the collision and the Turkish pilot (Lt. Halil Ibrahim Ozdemir) saved,

---

[23] Kathimerini, "Turkish jets chased off after violating Greek air space", http://www.ekathimerini.com/4dcgi/_w_articles_wsite1_1_20/08/2013_514765, accessed 18/09/2013, 20.12

after successfully ejecting safely, whilst eventually being picked up by a passing merchant ship, took place next to a passenger plane headed for Cairo. According to the Greek statement, "the planes collided at 8.000 metres (27.000 ft.) after the Turkish jets 'violated air traffic rules'. Greek military officials said the Turkish plane caused the collision by a 'sudden manoeuvre'"[24]. On the other hand, "the Turkish military reiterated that the crash had been caused by a Greek fighter interfering in Turkish manoeuvres in international airspace"[25]. Therefore, it is evident that the Aegean dispute is not only a theoretical issue; it has cost peoples' lives, interfering also with passenger planes, which fly over the Aegean Sea, a major concern for the Athens FIR, as will be evident below.

The heart of the airspace dispute is the FIR responsibility over the Aegean. In 1952 the Athens FIR and the Istanbul FIR were determined. "The limits of the Athens FIR were fixed at a point bordering the outer edge of the Turkish territorial waters leaving Greece in control of both international and Greek

---

[24] Smith, Helena and Ian Traynor, "Mid-air fighter plane collision risks new Greek-Turkish crisis", The Guardian, http://www.theguardian.com/world/2006/may/24/turkey.topstories3, accessed 02-10-2013, 13.10

[25] Ibid.

national airspace over the Aegean"[26]. Until 1974 no objections were raised; but after 1974, Turkey challenged the status quo. "With the issue of 'Notices of Airm-en' (NOTAM), Turkey tried to establish new 'Position Reports' for the Istanbul FIR in the middle of the Aegean Sea"[27], which seemed like a logical move by Turkey, since its provocative actions after invading and occupying Cyprus. In response to this, Greece issued NOTAM 1157/74 declaring the Aegean a danger area as is established by ICAO (International Civil Aviation Organization) rules, making it impossible to safeguard commercial planes passing through the Aegean. In 1980 Turkey cancelled its NOTAM, which resulted in Greece's removal of its NOTAM, returning to the status quo ante of the airspace over the Aegean.

Turkey understands the 10 mile airspace and especially the use of the Athens FIR as a Greek threat, by which Greece wishes to establish the Aegean as strictly a 'Greek Lake'. The Turkish Ministry of Foreign Affairs states that Greece uses the FIR responsibility it has "by demanding flight plans from the Turkish state"[28], whilst also accusing

---

[26] Syrigos, M. Angelos, *The Status of the Aegean Sea According to International Law*, (Athens, Bruxelles, Sakkoulas/Bruylant, 1997), p. 17-18

[27] Ibid., p.18

[28] Ministry of Foreign Affairs-Turkey, http://www.mfa.gov.tr /background-note-on-aegean-dispute.en.mfa, accessed 31-07-2010, 19.43

Turkey of violating the Athens FIR. In spite of these allegations it is crucial for Turkey to acknowledge that it should give flight plans, especially for its military planes, since over the Aegean Sea commercial planes fly on a daily basis and a crash is likely to take place, if precautions are not taken. All of this defiance from Turkey's part is a result of its fears and ambitions. Turkey would like the Athens and the Istanbul FIR to change by splitting the Aegean in to two parts. This will eventually give power to Turkey over the Aegean, which Greece is not likely to allow. It is a serious problem, due to the fact that any alteration to the current status quo will have negative results on the matter of sovereignty. In the case of splitting the Aegean into two parts, half of the Greek islands will be under the Istanbul FIR. Greek political and military planes will have to give flight plans to a foreign state to fly over Greek territory. This FIR dispute is seen by Greece as a direct threat to its sovereign rights, "fearing that any kind of responsibility given to Turkey to the west of Greek islands might later lead to further demands regarding their sovereignty"[29].

[29] Aydin, Mustafa, "Cacophony in the Aegean; Contemporary Turkish Greek Relations", University of Ankara, http://www.politics.ankara.edu.tr/eski/dosyalar/MMTY/27/6_mustafa_aydin.pdf, accessed 28/05/2010, 17.46, p. 119

D. Demilitarization and Militarization of the Eastern Aegean

A secondary issue, but still one of the disputed matters between Greece and Turkey, is the demilitarization of the Eastern Aegean Islands. The Treaty of Lausanne, Article 13, specifies that "no naval base and no fortification will be established"[30] in Mytilene, Chios, Samos and Ikaria, which are next to the Turkish coast. This does not only include these islands but also the Dodecanese islands, which were given to Greece by Italy in 1947 and also the Turkish islands Gokceada, Bozcaada and Rabbit Islands. However, the Treaty of Montreux (1936) allowed for the militarization of some islands near the Dardanelles. This is also a legal basis for Greece, which claims that the Treaty of Montreux cancelled the Treaty of Lausanne on this matter, meaning that the "demilitarized status of the islands"[31] was officially cancelled. The Treaty of Paris of 1947, through which the Dodecanese were turned over to Greece, refers also to a restriction of militarization; nonetheless, it does leave a small window where it accepts the existence of small forces for security

---

[30] Treaty of Lausanne, http://wwi.lib.byu.edu/index.php/Treaty_of_Lausanne, accessed 02/08/2010, 16.28

[31] Mann, Stephen, *The Greek-Turkish Dispute in the Aegean Sea: Its Ramification for NATO and the Prospects for Resolution*, (Monterey California, Naval Postgraduate School, 2001), p.28

purposes, which is manipulated by the two states to militarize both Greek and Turkish islands in the Aegean.

The reason why this matter has become one of the disputed issues in the quarrel is due to the fact that Greece has, since the 1960s, heavily militarized its Eastern Aegean islands. This is the result, as Greece claims, of the constant threats expressed by Turkish politicians and military against Greece and the Aegean status quo. These fears were confirmed when Turkey invaded Cyprus (1974). After 1974 Greece further fortified its islands, fearing that Turkey will also invade Greek islands in the Aegean. Greece's excuse for this action is based on Article 51 of the UN Charter which declares that a member state has the right of self defence. However, this has been seen by Turkey as an aggressive move towards its integrity and security. Also it is evident that Greece does have an advantage at this point, since its defensive systems and missiles can harm the Turkish mainland, as the islands are in close proximity to the Anatolian Peninsula. This reality resulted in Turkey's decision to create the 'Fourth Army', also known as the 'Aegean Army', which is situated in the South Western part of Turkey, opposite the Greek islands. The Turkish Ministry of Foreign Affairs explains that the Aegean Army is merely a "training army and has a defensive character"[32]. In spite of this statement it is vital to underline that this is an immediate threat towards Greece, since the 'Aegean Army' is the "largest non-oceangoing landing force in the world"[33] and is also not under NATO control.

The demilitarization and militarization of the Eastern Aegean Islands contributes to the general lack of trust between the two neighbouring states. As is evident, this case is a "chicken and egg situation"[34]. Both states claim that their armies are defensive and that they preserve the balance of power within the South Eastern part of the Aegean Sea. Although being true, in both cases, the mere presence of these two armies, opposite each other, do not facilitate in the disarmament of the two states. Due to this specific disputed issue between Greece and Turkey, NATO statistics verify that both spend an enormous part of their GDP on defence. This is one matter which has contributed towards the

[32] Ministry of Foreign Affairs-Turkey, http://www.mfa.gov.tr/background-note-on-aegean-dispute.en.mfa, accessed 31-07-2010, 19.43

[33] Mann, Stephen, *The Greek-Turkish Dispute in the Aegean Sea: Its Ramification for NATO and the Prospects for Resolution*, (Monterey California, Naval Postgraduate School, 2001), p.29

[34] Aydin, Mustafa, "Contemporary Turkish-Greek Relations: Constraints and Opportunities", in Aydin, Mustafa and Kostas, Ifantis (eds), *Turkish-Greek Relations, The Security Dilemma in the Aegean*, (London and New York, Routledge, 2004), p.30

recent economic crises in Greece that could change the regional stability, since the latter will not be able to continue its arms race with Turkey.

E. Disputed Islands, Islets and Rocks - Grey Zones

This is the last Aegean difference between Greece and Turkey and the most recent one. "This dispute is distinct from the other Aegean disagreements in that it constitutes the first, and in fact the only, dispute between Turkey and Greece involving territorial sovereignty"[35]. Although it seems that the disputed islands, islets and rocks are a different matter from the grey zones, they are actually interrelated.

The quarrel concerning the Imia (Greek name) / Kardak (Turkish name) rocks in 1996 was the beginning of arguments regarding the sovereignty of islands, islets and rocks. It all began in 1995 when a Turkish cargo ship ran aground the Imia rocks. The Turkish captain refused any assistance from the Greek authorities because he argued that he was in Turkish territory. This incident was later leaked to the Greek newspaper 'GAMMA'. The episode worsened when the mayor of Kalimnos, a Greek island near Imia, raised the Greek flag on the Imia

---

[35] Bolukbasi, Deniz, *Turkey and Greece The Aegean Disputes- A Unique Case in International Law*, (London, Cavendish Publishing Limited, 2004), p. 823

islet. Turkish journalists from the newspaper Hurriyet answered by lowering the Greek flag and raising the Turkish flag. That is when Greece, first, sent commandos and warships in the area and Turkey replied by doing the same. Both countries sent warships, planes and military personnel, which contributed to the tension build up in that area. Both were on the brink of war, when on the 31st of January 1996 the USA via President Bill Clinton mediated and achieved a stand down. However, this issue has yet to be solved, since both countries backed down and returned to the status quo ante. Both states still believe that the Imia / Kardak rocks are part of their sovereignty.

To understand this dispute we have to explain the location of the rocks and the treaties concerning it. Imia are in the Eastern Aegean. The Imia rocks are 3.6 miles from the Turkish coast and 5.5 miles from the nearest Greek island. Due to the proximity to Turkey the latter believes that they are Turkish. Also the Lausanne Treaty and the Treaty of Paris, by which the Dodecanese were given to Greece by Italy, do not name the Imia rocks. However, the Paris Peace treaty of 1947 clearly emphasises that Italy cedes to Greece 12 Dodecanese islands "specified by name as well as adjacent islets"[36]. Imia is one of the adjacent islets. Greece even contests any claim by

[36] Bolukbasi, Deniz, *Turkey and Greece The Aegean Disputes- A Unique Case in International Law*, (London, Cavendish Publishing Limited, 2004), p. 905

Turkey by basing its argument on the Lausanne Treaty "which left Turkey only those islands which are within the 3 mile limit"[37], excluding Kastelorizo in the South of the Anatolian Peninsula. Nonetheless, Turkey does base its claims on other treaties which Greece does not accept, and vice versa. Additionally, Turkey asserts that Greece approached Turkey in 1950 and 1953 to verify previous Treaties concerning the Dodecanese and the islets. Greece refuses that anything like this ever took place. It seems that both states are playing a 'cat and mouse' game where no one seems to be winning, where the official goal of both is to dispute the other on any matter, in relation to the islands. Even so, a great fact that should not be disregarded is that the Turkish ambassador in Rome warned Prime Minister Tansu Ciller before and during the Imia crises that "Italy obtained documents which proved without any doubt that the islet of Imia was part of the Dodecanese and was indisputably Greek"[38]. Also the 1932 Protocol between Turkey and Italy established

---

[37] Acer, Yucel, *The Aegean Maritime Disputes and International Law*, (Burlington, Ashgate Publishing Limited, 2003), p. 22

[38] "...η Ιταλία κατείχε έγγραφα τα οποία αποδεικνύουν πέραν πάσης αμφιβολίας ότι η νησίδα των Ιμίων ανήκε στα Δωδεκάνησα και ήταν επομένως αναμφισβήτητα ελληνική.", Ellis, Athanasios and Michalis, Ignatiou, *IMIA- TA APORRITA TILEGRAFIMATA TON AMERIKANON*, (Athens, Livani, 2009), p. 17

that Imia was then Italian, which means that after 1947 they became undoubtedly Greek, as there was a median line which passed "between the Imia islets on the Italian side and Kato Island on the Turkish side"[39].

Through this incident Turkey, in 1996, started promoting the idea of grey zones in the Aegean. Turkey supports the view that no Treaty has legally resolved the sovereignty issue of many islands, islets and rocks, since they are not named in any Treaty. These areas are called grey zones, meaning that they are "territories without sovereignty and are therefore stateless"[40]. However, these areas have yet to be named. The only determination in regards to the so called grey zones were given to the public by an article, published by Milliyet, a Turkish daily newspaper, but no official answer has been given to Greece in regard to the areas in question.

Greece on the other hand states that numerous islands have not been mentioned, but that does not mean that they are disputed, or stateless. Greece has expressed the view that the Treaty of

---

[39] Athanasopoulos, Haralambos, *Greece, Turkey and the Aegean Sea*, (Jefferson, North Carolina, McFarland and Company Inc. Publishers, 2000), p. 76

[40] Paulenoff, Alex, Perspectives on Global Issues, "The Aegean Sea Continental Shelf Dispute: Greek and Turkish Interpretations", http://www.perspectivesonglobalissues.com/archives/spring-2009/continental, accessed 23/05/2010, 7.19 p.m.

Lausanne (Article 12) clearly affirms that any islands, islets and rocks within 3 miles of the Turkish coast belongs to Turkey (except for Kastelorizo, which is 2.5 miles of the Turkish coast and belongs to Greece). The rest of the islets and rocks in the Aegean outside the 3 miles from the Anatolian peninsula belong to Greece. Conversely, this view is contested by Turkey, believing that not every islet and rock belong to Greece but to the successor of the Ottoman Empire, insinuating Turkey. Ahmet Davutoglu also explains that "the fact that the overwhelming majority of the islands in the Aegean are under Greek sovereignty is the most important political stalemate for the adjacent maritime region of Turkey"[41] Therefore, it is apparent that the difference in interpretation of the treaties is the main problem concerning this disputed issue. Yet again this is understood in Athens as a way of bullying Greece in to negotiating the current status quo in the Aegean. Nonetheless, it seems that the issue relating to the sovereignty rights of the Aegean Islands "will prove the hardest to settle"[42]

---

[41] «Το γεγονός ότι η συντριπτική πλειονότητα των νησιών του Αιγαίου βρίσκεται υπό ελληνική κυριαρχία αποτελέί το σημαντικότερο αδιέξοδο της πολιτικής της εγγύς θαλάσσιας περιοχής της Τουρκίας».
Davutoglu, Ahmet, *To Stratigiko Vathos, I Diethni Thesis tis Tourkias,* (Athens, Poiotita, 2010), p.268

[42] Prescott, J.R.V., *The Maritime Political Boundaries of the World,* (London and New York, Methuen & Co, 1985), p. 310

Conclusion

Greece and Turkey, two ally states, are in a chronic dispute over many issues concerning the Aegean Sea. It is important to clarify why they are in dispute over the Aegean, which is the main bilateral problem between them, but not the only quandary between the two states. I will not go into detail, since the main theme here is the Aegean Sea dispute, but it is vital to realise that other problematic matters have facilitated into the crisis build up over the Aegean. The two key issues are the Cyprus invasion in 1974 and the minority issues within the two states. Specifically, the Cyprus problem has produced many bilateral problems between Greece and Turkey, such as the militarized state of the Eastern Greek islands and the establishment of the Aegean Army.

The Aegean dispute is a challenge for both states, consisting of many difficulties. The reason why this persists is evident when one takes into account the oil and natural gas resources available in the seabed. Whichever state obtains the maximum possible control in the Aegean will inevitably boost its economy. Additionally, due to the fact that the Aegean Sea connects the Mediterranean Sea with the Black Sea, it has a strategic importance for the state that controls it. Politically it is a theme used by both countries to help with elections. Culturally and historically it is evident that the Aegean was always a 'Greek Lake', but in the modern globalised world it is important for Greece to understand that Turkey also has rights in the Aegean. However, the latter should stop promoting new problems concerning the Aegean dispute, such as the grey zones. From the above analysis it is apparent that "Greece is not the demandeur in most of the issues of the Aegean".[43]

[43] Heraclidis Alexis, "The Greek-Turkish Conflict: Towards Resolution and Reconciliation", in Aydin, Mustafa and Kostas, Ifantis (eds), *Turkish-Greek Relations, The Security Dilemma in the Aegean*, (London and New York, Routledge, 2004), p.82

CHAPTER 2
The Consequences of the Aegean Dispute for NATO

Introduction: The Two States in NATO

The Aegean Dispute between Greece and Turkey does not only have consequences for the two neighbouring states, on a bilateral level, but it has also affected international bodies, one of them being NATO. All the aspects of this dispute (continental shelf, territorial sea, airspace, demilitarization of the Eastern Aegean islands, disputed islands, islets, rocks and the grey zones) have irritated the alliance in many occasions during military practices, especially when they take place in the Aegean Sea. This is the result of 'NATOizing' the Aegean dispute, to help each state, respectively, with its claims in the region.

Both states entered the alliance in 1952, which was a great benefit for NATO since it "was intended to harness the two neighbours to the strategic goals of the West"[44] making it possible to organize its defence systems during the Cold War era in the Southern part of Europe against the Soviet Union. Since 1974 and the invasion of Cyprus by Turkey this

[44] Chircop, Aldo, Andre, Gerolymatos and John, Iatrides (Eds), *The Aegean Sea after the Cold War- Security and Law of the Sea Issues,* (London. Macmillan Press LTD., 2000), p. 240

has changed. The headquarters of NATO were established in Izmir, Turkey, under American General and Greek and Turkish Deputies. When "Greece withdrew from the military wing of NATO in 1974"[45], because of the Cyprus invasion, a Turkish General was appointed in charge of the headquarters of NATO in Izmir, in place of the American General. Due to the invasion in Cyprus the USA placed an embargo on Turkey. Relations between the three countries were not sound. In 1975 Greece applied to re-enter the alliance and talks commenced on the establishment of a Greek-NATO headquarters in Larissa, Greece. The reason for re-entry into NATO was due to Karamanlis', Greek Prime Minister at the time, "concern that Turkey would further advance its claims in the Aegean, with Greece's absence from the military command"[46]. This move was vetoed by Turkey, which wanted to determine the operational boundaries of the new headquarters, by expanding the Turkish control over the Aegean and consequently over some of the Greek islands. This would have produced great disputes over the airspace and FIR control over the Aegean. However, Greece pushed for the status quo ante. Due to various threats by the USSR and Iran the delimitation

---

[45] Syrigos, M. Angelos, *The Status of the Aegean Sea According to International Law*, (Athens, Bruxelles, Sakkoulas / Bruylant, 1997), p. 21

[46] Moustakis, Fotios, *The Greek - Turkish Relationship and NATO*, (London, Frank Cass Publishers, 2003), p. 36

of the operational boundaries of the two headquarters were not resolved before the reintegration of Greece into the alliance. This topic has yet to be resolved, contributing to the instability of Greek-Turkish relations.

Both Greece and Turkey have vetoed, over the years, the other state's NATO funding, concerning the defence systems and the headquarters in Izmir and Larissa. This caused problems for NATO, making modernization within the alliance a complicated issue to solve. Projects and missions have been halted due to this practice of vetoing each other. Even in the post-Cold war era, were NATO does not have a clear enemy as it did during the Cold war epoch, it seems that it is being threatened from within the alliance through Greece's and Turkey's disputed issues, which inevitably affect NATO's cohesion. An evident result of the mistrust of both countries is the arms race. "In total disregard of NATO strategy, the defence doctrine of Greece is focused entirely on the perceived Turkish threat and extends Greece's military responsibilities to Cyprus, thus increasing dangerously the risk of unintended clashes with Turkey"[47]. On the other hand Turkey allocates part of its NATO forces in Northern Cyprus. The importance of this arms race is understood by

[47] Iatrides, O. John, "NATO and Aegean Disputes: The Cold War and After", in Chircop, Aldo, Andre, Gerolymatos and John, Iatrides (Eds), *The Aegean Sea after the Cold War-Security and Law of the Sea Issues*, (London. Macmillan Press LTD., 2000), p. 36

the two allied states and by NATO in respect to the stability or instability of the region especially when we take into account the fact that "by the end of 1992 Greece and Turkey were the greatest importers of military material worldwide"[48]. Both states feel threatened by the other, continuing their increase in military weaponry, buying more weapons than they need. In 2009 Greece purchased over $6 billion worth of weapons, making Greece the 5th largest arms buyer in the world. These facts in regard to the arms race of both countries explain how they are "basically arming to fight one another"[49], which would have a negative affect for NATO. The consequences for NATO are its vulnerability in its South-Eastern flank, whilst the decision process highlights the difficult atmosphere which prevails due to the dispute over the Aegean. This is also pointed out especially when military exercises are taking place within the Aegean area, whilst NATO maintains its neutral stance towards both states when a crisis occurs. Bellow it will be evident,

[48]Tsakonas, Panayiotis, Weatherhead Center for International Affairs, " Greek-Turkish Relations Towards the 21st Century- A view from Athens", http://www.wcfia.harvard.edu/ fellows/papers/1998-99/tsakonas.pdf , accessed 30/05/2010, 16.10

[49] Mann, Stephen, *The Greek-Turkish Dispute in the Aegean Sea: Its Ramification for NATO and the Prospects for Resolution*, (Monterey California, Naval Postgraduate School, 2001), p.52

through examples and case studies from past incidents, how this atmosphere of mistrust and aggressiveness is still the case today between the two states.

A. NATO's Vulnerability in its South-Eastern Flank

The Aegean dispute between Greece and Turkey does not only affect them, on a bilateral level. It also promotes regional instability. This has had, inevitably, consequences for NATO. This was a serious problem during the Cold war era, when the alliance had to be ready against a certain enemy, i.e. the USSR. However, the bilateral problems did not affect the alliance during the Cold war period since both Greece and Turkey had to face a common enemy, the Warsaw Pact[50]. After the Cold war both states saw the other as the enemy in a more aggressive manner; that is why, even today, in the post-Cold war period the Aegean dispute and any bilateral crises in Greek-Turkish relations are crucial concerns. Therefore the alliance's presence in the region is still vital, as was apparent in the war in

---

[50] The Warsaw Pact (1955) was a treaty signed by the USSR and seven of its European satellites, where a mutual organisation was established, giving the Soviets command over the armed forces of the member states, which included the USSR, Albania, Poland, Hungary, Romania, Czechoslovakia, Bulgaria and East Germany.

Yugoslavia and during the wars in Iraq and Afghanistan.

The vulnerability of NATO in its South-Eastern flank is the most serious consequence of this Aegean dispute between the two countries for the alliance, since it does not only concern Greece and Turkey, but it also influences the stability of NATO, hence it concerns all the states within the alliance, who distinguish the weak spot of the coalition. A major fear for NATO is a conflict between Greece and Turkey which "could lead to the collapse of NATO's Southern Flank and would severely disrupt the Alliance's efforts to play a meaningful role in the post-Cold War world"[51]. This was the case when both Greece and Turkey came close to war in 1996 over the Imia rocks in the Eastern Aegean. This occurrence had made evident not only NATO's neutrality but also its inability to act and to prevent any conflict within its alliance. Nevertheless, it is significant to identify that it would be difficult for NATO to take sides if a conflict between two of its members occurred; on the other hand, this does not explain its incapability in preventing it.

NATO's vulnerability within that region is evident through the military exercises which have taken place in the Aegean Sea. One of the most recent incidents was in 2000 where a NATO exercise

[51] Dokos, P. Thanos, "Tension-Reduction and Confidence-Building in the Aegean", in Aydin, Mustafa and Kostas, Ifantis (eds), *Turkish-Greek Relations, The Security Dilemma in the Aegean*, (London and New York, Routledge, 2004), p.125

named 'Destined Glory' took place in the Southern region of the alliance, which was the largest NATO exercise since the endorsement of the new command structure in 1999. "National disagreements between Greece and Turkey concerning airspace management eventually resulted in Greece withdrawing its military participation from the last series of the exercise"[52]. This was the result of the Aegean dispute over the demilitarization of the Eastern Aegean islands. Greece used air corridors, which were approved by NATO for this exercise, over the islands of Lemnos and Ikaria, which are militarized islands. Turkey perceived this as "a violation of their demilitarized status"[53] thus making this unimaginable for Turkey to allow. Hence it sent Turkish fighter planes to intercept the Greek planes in the middle of the planned exercise. On the other hand, in response to this, "anti-aircraft missiles locked on to Turkish planes that flew close to the Greek-controlled part of Cyprus"[54]. This incident is a serious issue for NATO,

---

[52] Global Security, "Destined Glory", http: //www.globalsecurity.org/military/ops/destined-glory.htm, accessed 13/07/2010, 16.01

[53] Ministry of Foreign Affairs - Turkey, http://www.mfa.tr/the-_destined-glory-2000_-exercise_br_no_193--_october-23_-2000.en.mfa, accessed 28/06/2010, 15.37

[54] Mann, Stephen, *The Greek-Turkish Dispute in the Aegean Sea: Its Ramification for NATO and the Prospects for Resolution*, (Monterey California, Naval Postgraduate School, 2001), p.51

since part of the territory of a member state is not allowed to be used in military exercises due to disputed matters with a fellow member. This could prove a threat and a weakness in any future combat situation. It emphasises the vulnerable circumstance which prevails in the South-Eastern part of the alliance. This could prove to be not only a regional problem for NATO but also a vital predicament for the existence of the alliance in the future, if war occurs near Greece and Turkey.

The 'Destined Glory' exercise of 2000 is not the only case where the island of Lemnos and the use of its airspace have provoked the two sides. Greece has repeatedly boycotted military exercises when Lemnos is excluded from these exercises. An example of this is the 1983 NATO exercise where Lemnos was not included, to pacify Turkey, but was part of the "free play part"[55], which quietened down Greece. The militarized status of the island and its dominant geographical position, very close to the Dardanelles, has constantly stopped or undermined military exercises in that region. Statements such as "Athens says Turkey has behaved unacceptably by closing its air space to Greek aircraft and harassing its pilots"[56] should not prevail in NATO exercises.

---

[55] Ibid., p.47

[56] Morris, Chris, "Analysis: Hostility returns to Aegean", BBC news, http://news.bbc.co.uk/1/hi/world/europe/984135.stm, accessed 04/08/2010, 20.14

Ankara has achieved to make the Lemnos issue, and consequently the demilitarization dispute with Greece, a NATO problem by preventing a NATO island to not take part in military exercises. Accounts like this continue the clash between Greece and Turkey and explain why these two neighbouring states are still in dispute over the Aegean Sea.

"The tension between these two allied countries undermines Alliance harmony and has a deliberating effect on the security of NATO's Southern Flank".[57] This tension between the two states is the main and direct threat to the South-Eastern flank of NATO. It is expressed by the undermining of the military exercises in the Aegean and the Eastern Mediterranean which "affects the readiness, cohesiveness and confidence of the forces and, therefore, of the alliance"[58]. However, it is imperative that a solution should be found soon, not only on a bilateral level but also on an international one. Being members of NATO and other international organizations, both Greece and Turkey should take into account the defence and

---

[57] Karaosmanoglu, Ali, " Turkey and the Southern Flank: Domestic and External Contexts", in Chipman, John (Eds.), *NATO's Southern Allies: Internal and External Challenges,* (London and New York, Routledge, 1988), p. 338-339

[58] Mann, Stephen, *The Greek-Turkish Dispute in the Aegean Sea: Its Ramification for NATO and the Prospects for Resolution*, (Monterey California, Naval Postgraduate School, 2001), p.47

maintenance of the South-Eastern part of the alliance, since beyond but not far from its borders war still prevails, i.e. in Iraq and Afghanistan, whilst the Arab Spring has created new issues for the Eastern Mediterranean and the Middle East. The Aegean dispute should not be a reason for the failure of defending the alliance and for not going ahead with military exercises over the Aegean, which is a part of NATO. The alliance's readiness, from a military point of view, is constantly compromised. "At its best it confounds, confuses and hinders NATO operations; at its worst, in the event of actual hostilities between the two allies, it could break up or split the alliance"[59]. Although the Aegean dispute and its consequences for NATO appear to have a secondary importance for the alliance, since its ventures in Yugoslavia during the last decade of the 20th century and the wars in Iraq and Afghanistan during the first decade of the 21st century seem to be of greater significance; nonetheless, it still provokes chronic implications for NATO, which is not only obliged to fight external threats but also has to handle an internal risk.

B. Problems in Taking Decisions

The second consequence is a direct result of the Aegean Dispute and generally of the Greek - Turkish relations. NATO faced a barrier when

---

[59] Ibid., p.52

decisions had to be taken and unanimous endorsements had to be reached. This was made manifest in two cases where the disputes between Greece and Turkey prevailed in NATO's efforts to decide on policies.

The first instance is from 1981, when Greek Prime Minister Andreas Papandreou asked NATO if it will assure Greece that Article 5 will be enforced if a threat occurs from outside or inside the alliance. This article states that "The Parties agree that an armed attack against one or more of them in Europe or North America shall be considered an attack against them all"[60], meaning that military action would be taken to preserve the status quo of the alliance. This was a clear effort to bring the bilateral problems and threats to the alliance, which was not accepted by Turkey and consequently meant Greece's defeat on this matter. Many NATO members answered by quoting Article 1 of the North Atlantic Treaty which clearly emphasises that each state should "settle any international dispute in which they may be involved by peaceful means..."[61] The final statement was not approved by the Greek Prime Minister. This was the first time in the history of the alliance that a communiqué was not approved, since it required the unanimous consent.

---

[60]The North Atlantic Treaty, http://www.nato.int/cps/en/natolive/official_texts_17120.htm, accessed 05/08/2010, 16.58

[61] Ibid

The second case was in 1992 when the "NATO Command Structure Document MC-294"[62] was granted by the NATO Military Committee but was not ultimately sanctioned by the NAC (the North Atlantic Council). The body in charge of political decisions could not approve any unanimous verdicts due to the fact that Greek and Turkish politicians did not agree on certain issues concerning the militarization status, the method with which the forces were arranged and many more. Concisely, it is understood that both states continued their dispute in NATO, dissatisfying the other members. This specific incident did not allow for the changes planned by the MC-294 to take effect. Finally in 1999 the new Command Structure was agreed. "The Greek - Turkish dispute caused NATO over seven years of not receiving the benefits and improvements that MC-294 would have brought earlier"[63].

Although NATO, through the ratification of the new command structure achieved the capability of taking decisions without the need for a unanimous vote, it also worsened the situation in the Aegean,

---

[62] Mann, Stephen, *The Greek-Turkish Dispute in the Aegean Sea: Its Ramification for NATO and the Prospects for Resolution*, (Monterey California, Naval Postgraduate School, 2001), p.48

[63] Mann, Stephen, *The Greek-Turkish Dispute in the Aegean Sea: Its Ramification for NATO and the Prospects for Resolution*, (Monterey California, Naval Postgraduate School, 2001), p.49

which inevitably means that all the consequences stated in this book still prevail. NATO achieved this by placing "the entire Aegean airspace under its operational control and command"[64]. Under NATO's umbrella Turkish fighter planes fly over the Aegean, violating Greek airspace, provoking problems for the advancement of Greek – Turkish relations.

These case studies emphasise that the Greek – Turkish disputes and relations have affected the alliance not only on the operational level, by holding back its capabilities of improvement, but it has also brought friction to its cohesion, affecting the other members, since improvement was halted. Some of the problems (for example both states "vetoing each other in NATO fora"[65]), through the 1999 new command structure, have been limited, but inconveniences still exist due to the Aegean dispute.

C. NATO's Neutral Stance

This third consequence of the Aegean dispute for NATO has also consequences for Greece and

---

[64] Athanasopoulos, Haralambos, *Greece, Turkey and the Aegean Sea*, (Jefferson, North Carolina, McFarland and Company Inc. Publishers, 2000), p. 103

[65] Tsakonas, Panayiotis, Weatherhead Center for International Affairs, "Greek-Turkish Relations Towards the 21st Century- A View from Athens", http://www.wcfia.harvard.edu/fellows/papers/1998-99/tsakonas.pdf, accessed 30/05/2010, 16.10

Turkey who perceive the alliance in a certain way, especially when both countries come against each other. In any case it is understandable that a conflict between the two NATO members would leave the alliance in a difficult position, without being able to take a side. The only solution for NATO would be to try and prevent a conflict.

NATO has been tested and been evidently neutral when Greek – Turkish relations deteriorate. The principal examples would be the Cyprus invasion in 1974 and the Imia crises of 1996, when both came close to the brink of war. But it is also a fact that "neither Greece nor Turkey believes that NATO is truly impartial; both believe that NATO favours the other"[66]. Greece considers that NATO is biased towards Turkey, due to its strategic significance, its bases and its proximity to the Middle East, especially in recent years, with the wars in Iraq and Afghanistan and the Arab Spring, which has spread across North Africa and the Middle East. On the other hand, Turkey believes that NATO is prejudice towards Greece in view of the fact that the latter has cultural and historical relations with the USA and Europe, whilst its strategic advantage is apparent. These beliefs from both countries, together with the neutral stance that NATO has taken towards the

[66] Mann, Stephen, *The Greek-Turkish Dispute in the Aegean Sea: Its Ramification for NATO and the Prospects for Resolution*, (Monterey California, Naval Postgraduate School, 2001), p.52

Aegean dispute and generally the Greek-Turkish relations, highlight the poor alliance unity.

This neutral stance, as a consequence of the Aegean dispute, has left other international powers to intervene and try to solve the problem between Greece and Turkey. During the last decades "the United States was successful in its crisis management role, which preserved the stability of NATO in South-Eastern Europe by avoiding the unthinkable: the two allies going to war"[67], emphasising the fact that the USA has great political and military power and presence within the Balkan and the Aegean region. The neutral stance is also emphasized by John Iatrides who explains "if NATO is helping to 'keep the peace between Greece and Turkey' beyond purely American diplomatic efforts, the alliance is doing so in mysterious ways"[68]. However, it appears that the alliance is interested in a resolution of the problematic relations of both allied states as expressed by the Secretary General of NATO, Dr. Javier Solana when he stated that "in the 1988 Memorandum of Understanding both countries

[67] Kalaitzaki, Theodora, "US Mediation in Greek-Turkish Disputes since 1954", *Mediterranean Quarterly* 16.2 (2005), p. 123

[68] Iatrides, O John, "NATO and Aegean Disputes: The Cold War and After", in Chircop, Aldo, Andre, Gerolymatos and John, Iatrides (Eds), *The Aegean Sea after the Cold War-Security and Law of the Sea Issues*, (London. Macmillan Press LTD., 2000), p. 36

(Greece and Turkey) have recognized the obligation to respect the sovereignty and the territorial integrity of each other"[69]. This stresses that NATO does want to promote peace and stability between the two states, but does not in any case seem persistent in acting to achieve this goal. The Imia crises in 1996 depicts emphatically that only the USA, which mediated through President Bill Clinton, achieved the prevention of war.

Conclusion

Through the examination of the consequences of the Aegean dispute between Greece and Turkey for NATO it is evident that they can only be described as challenging and agonizing for the alliance. Then again the alliance does have other, more important issues to solve; nevertheless, the Aegean dispute undermines NATO's integrity from within. Any conflict within the alliance "would have been devastating for NATO, whose new role includes peace implementation and peacekeeping, in addition to its more traditional role of collective defence"[70].

---

[69] NATO, "Statement by the Secretary General of NATO, Dr. Javier Solana, on Confidence Building Measures Between Greece and Turkey", http://www.nato.int/cps/en/natolive/news_25854.htm?selectedLocale=en, accessed 04/08/2010, 21.36

[70] Bahcheli, Tozun, Theodore A., Couloumbis and Patricia, Carley, *Greek-Turkish Relations and U.S. Foreign Policy -*

The fact that both states are members of NATO and their bilateral difficult relations have consequences not only for the alliance but also for the other member states, is an important reason for a rapid resolution of all the differences that have prevailed these past decades. "Many attempts at resolution have been made, with few successes"[71]. One example of this challenge is the 1999 new command structure, which inevitably came to existence so the alliance can function and continue its works. Greece on the 1st September 1999 established a NATO base in Larissa, which was under a Greek General. The new centre shared responsibilities for parts of South-Eastern Europe. However, this base was closed down on the 21st of June 2013. The closing down of the headquarters in Greece was enforced due to the framework of the new command structure of NATO, which was decided at the 2010 Lisbon Summit[72]. However,

---

Cyprus, the Aegean and Regional Stability, (Washington, United States Institute of Peace, 1997), p. 7

[71] Mann, Stephen, *The Greek-Turkish Dispute in the Aegean Sea: Its Ramification for NATO and the Prospects for Resolution*, (Monterey California, Naval Postgraduate School, 2001), p. 53

[72] There we find, in article 49, the following statement on the introduction of a new NATO Command Structure: "We have agreed a framework for a new NATO Command Structure, which will be more effective, leaner and affordable. It will also be more agile, flexible, and better able to deploy on operations, including Article 5 contingencies and providing

these negative consequences should not prevent future initiatives towards a peaceful and effective resolution of the Aegean dispute. If a solution is not found then all the negative consequences this dispute has for the alliance will continue to exist. Perhaps NATO can take this troublesome relationship concerning the Aegean dispute between its two members and resolve any challenging matters that prevail in their bilateral relations. This would be a positive action, which will also benefit NATO's integrity and continuity. On the other hand, it is possible that problems and disappointments might occur if a final solution is found. Even so, if NATO does not have the power and the will to solve these issues, then maybe both states should look towards another international body which could, at least from a political point of view, solve the Aegean dispute, hence solving the negative consequences for NATO. This body could be the European Union,

---

visible assurance. The new structure represents a significant reduction in the number of headquarters and a manpower saving of 35%, representing almost 5,000 posts, or more, if and where possible. It will have a new relationship with our national headquarters, and will also ensure a regional focus. A final decision on a new NATO Command Structure, including its geographic footprint, will be taken no later than June 2011, and we have tasked the Secretary General to prepare proposals to this end".
NATO, "Lisbon Summit Declaration", http://www.nato.int/cps/en/natolive/official_texts_68828.htm?mode=pressreleas, accessed 01/10/2013, 09.30

which will be discussed and analysed in the next chapter.

CHAPTER 3
The consequences of the Aegean Dispute for the EU

Introduction

The consequences of the Aegean dispute between Greece and Turkey for the European Union will be discussed in this chapter. Greece is a full member state and Turkey aspires to become a member of this Union. To understand why this bilateral issue affects the EU it is vital to present a brief history of the two nations with respect to the Union. Then it will be evident why and how the Aegean dispute has consequences for the EU.

Both states applied in 1959 to become members of the European Economic Community (EEC). However, due to difficulties in both cases (various political upsets and later the junta dictatorship in Greece, which lasted from 1967 until 1974 and the difficulties with democratic institutions in Turkey and economic instability in both states) it made it impossible for them to become members at that time. Eventually, Greece did become a member on the 1st of January 1981. Due to this accession, Turkey has found it harder to become a member, since Greece has been vetoing its accession for years by blocking any "EU aid and financial cooperation with Turkey"[73], due to their disputed issues, not only in the Aegean but also in Cyprus.

This all changed in 1999 at the Helsinki Summit, where Greece changed its policy concerning Turkey and gave the chance to the latter to achieve a candidate state status. If Greece had continued vetoing Turkey's accession it would probably had found itself further isolated within the Union. However, Greece saw positive results arising from its policy change towards Turkey. In the following years talks commenced for Turkey's full accession within the EU. As explained by Scott Bennett "countries that face enduring disputes have more possibilities of resolving those disputes during a period in which both countries happen to be democratic"[74]. It is a fact that Greece is a full democratic state. Turkey, by becoming a candidate and eventually a member of the EU will mean that it has already achieved a full democratic status. As Doyle believes, due to the expansion and maintenance of liberal, democratic states "the prospects for a world peace"[75] have been

---

[73] Barchard, David, "Turkey and the European Union", Centre for European Reform, http://www.cer.org.uk/pdf/p093_turkey.pdf, accessed 07/08/2010, 16.55

[74] Patsiaouras, Konstantinos, *Democratic Peace theory and Greek-Turkish Relations in the Context of the European Union*, (California, Thesis Naval Postgraduate School, 2009), p. 27

[75] Doyle, Michael W., 'Kant, Liberal Legacies and Foreign Affairs', *Philosophy and Public Affairs,* Vol. 12, No. 3, (Summer, 1983), p. 206

strengthened. It is evident that when Turkey becomes an official member of the European Union and it follows Bennett's theory, then disputes between Greece and Turkey will be gradually solved. However, in 2004 the European Council came to a decision to reassess Turkey's progress and in 2006 the EU suspended further talks with Turkey, promoting the idea of a special relationship between Turkey and the EU, which the Mediterranean state could not accept. Turkey has yet to become a full member of the EU and it seems like Turkey is actually "turning to the East"[76], towards the Arab states, leaving behind the European Union and the difficulties found during its accession period.

Despite any eventual result in Turkey's accession to the EU it is evident that, especially since Greece's acceptance within the union, disputes between the two Aegean states have brought negative consequences to Europe. The union seems that it is a place for both Greece and Turkey to resolve their bilateral problems. This action disappoints the other member states, which up to this day have taken no action to solving the Aegean and any dispute between the two states. The consequences for the EU due to the Aegean dispute are the problems relating to the prevention of illegal immigrants coming from Turkey to the European Union, through Greece, and the difficulties for

[76] Turgut, Pelin, 'Turning to the East', Time Magazine, (July 5, 2010), p. 26

FRONTEX[77] to achieve its goals in the Aegean. Also the regional stability is compromised since the Union's borders in the Aegean, and hence Greece's borders, have not been delimitated, which preserves the dispute and gives Turkey the power to continue disputing European territory, making it difficult to police the area, which also relates to the first consequence. Finally, the EU's role in this dispute, especially its neutral position, is emphasized, thus showing its disunity, as was evident in NATO's case.

A. Immigration. FRONTEX

The biggest consequence of the Aegean dispute for the European Union is the illegal immigration, which has in the past few years boomed in the Aegean Sea. The Greek islands receive immigrants that actually come to them through

[77] "Frontex promotes, coordinates and develops European border management in line with the EU fundamental rights charter applying the concept of Intergrated Border Management. Frontex helps border authorities from different EU countries work together. Frontex's full title is the European Agency for the Management of Operational Cooperation at the External Borders of the Member States of the European Union. The agency was set up in 2004 to reinforce and streamline cooperation between national border authorities. In pursuit of this goal, Frontex has several operational areas which are difined in the founding Frontex Regulation and a subsequent amendment..."
Frontex, http://frontex.europa.eu/about-frontex/mission-and-tasks, accessed 31/01/14, 15.28

Turkey, making this a significant problem for both Greece and the EU. Statistics show that from 1998 until 2002, "almost two-thirds of the incidents associated with the trafficking or transport of illegal migrants in the Mediterranean Basin were between Turkey and Greece"[78]. In 2009 more than 150.000 illegal migrants entered the EU through Greece, mainly due to the fact that other routes of entry to the EU have been safely guarded or closed (i.e. Spain, Italy, and Malta).

This is evidently a principal worry for the EU, since it has ratified a number of laws concerning illegal immigration from outside the Union. The problem relates to the Aegean dispute, due to the fact that it is a negative consequence of the bilateral issues between Greece and Turkey. Moves have been made for talks to be commenced between the Greek and Turkish governments, but the immigration business keeps prevailing, as it is profitable. The EU has initiated a "Stability Pact for Southeastern Europe"[79] and has established mechanisms for reinforcing border security and policing. Additionally, both Greece and Turkey have signed numerous agreements, especially after 1999 where Greece changed its stance towards Turkey's candidacy to the

[78] Rumelili, Bahar, 'The European Union's Impact on the Greek-Turkish Conflict', *The European Union and Border Conflicts: The Impact of Integration and Association (EU BorderConf),* No.6 (January 2004), p. 24

[79] Ibid., p. 24

European Union, on various subjects, one being the issue of "illegal immigration"[80]. Nevertheless, it should be pointed out that Turkey's ambitions of becoming an EU member will be the result of fulfilling an agenda given by the EU. This evidently proposes that its relations with both Greece and Cyprus have to be taken in to account; especially the disputes with Greece and the immigration challenges have to be rapidly resolved.

This specific matter also relates to the Aegean dispute due to the fact that Turkey interferes with EU patrols concerning immigration, mainly FRONTEX, which secures the external borders of the European Union and has never been accepted by Turkey. "One main reason was that the FRONTEX patrols take place at the boundaries of the Athens FIR, which translates, in practice, to recognize these borders and boundaries as Europe's external borders"[81]. This is a direct consequence of the Aegean dispute,

---

[80] Migdalovitz, Carol, "Turkey: Issues for U.S. Policy", Congressional Research Service, http://www.fas.org/man/crs/RL31429.pdf, accessed 20/05/2010, 15.26, p. 14

[81] "Ένας βασικός λόγος ήταν το γεγονός ότι οι περιπολίες της FRONTEX γίνονται στα όρια του FIR Αθηνών, κάτι που μεταφράζεται, πρακτικά, σε αναγνώριση αυτών των ορίων ως εξωτερικών συνόρων της Ευρώπης", Antoniou, Dora, "Epikindines Tourkikes prokliseis sto Aigaio, Apo tis parenohlisis aeroskafon tis FRONTEX meso rantar, stis epihirisis anaxaitisis elikopterou tou Limenikou apo F-16", *KATHIMERINI,* http://news.kathimerini.gr/4dcgi/_w_articles_2_ 28/03/2010_395803 , accessed 28/03/10, 14.56

making it difficult for an EU institution to achieve its goals, and it also intensifies the current situation over the Aegean. However, one incident in 2010 involved two Turkish F-16s which "flew alongside the aircraft from the FRONTEX force"[82]. The two F-16s were intercepted by two Greek fighter planes, allowing FRONTEX to continue its mission. The Greek Ministry of Defence claimed that this took place over the Greek island of Farmakonisi, which is part of the 'grey zones', according to Turkey, a reality not accepted by Greece. Ankara attempts to "consolidate its disputed problems in the Aegean"[83], through negotiation and cooperation between Turkey and FRONTEX, which would bring to an end the current situation between the two and will battle illegal immigration; however, the above case shows that Turkey has yet to establish a satisfactory agreement in regards to this problem. It is evident that the EU and FRONTEX are in a way used by both Greece and Turkey to achieve a resolution of the Aegean dispute in each state's favour. On the other hand, this is a

---

[82] Hurriyet Daily News, "Greece Says Turkish F-16s Harass EU Immigration Patrols" http://www.hurriyetdailynews.com/n.ph?n=greece-says-turkish-f-16s-harass-eu-immigration-patrols-2010-03-19, accessed 19/07/2010, 16.48

[83] "...να παγιώσει τις αμφισβητήσεις της στο Αιγαίο", Antoniou, Dora, "Epikindines Tourkikes prokliseis sto Aigaio, Apo tis parenohlisis aeroskafon tis FRONTEX meso rantar, stis epihirisis anaxaitisis elikopterou tou Limenikou apo F-16", *KATHIMERINI,* http://news.kathimerini.gr/4dcgi/_w_articles_2_28/03/2010_395803 , accessed 28/03/10, 14.56

problem for the European Union, since institutions like FRONTEX find it hard to succeed in achieving their goals; one case being the prevention of illegal immigrants coming through the Aegean Sea from Turkey. Nevertheless, this issue could be resolved if or when Turkey becomes an EU member, since EU borders will no longer be a quandary for Turkey (but this does not mean that problems concerning the borders with Greece will not continue), hence halting its dangerous manoeuvres with the FRONTEX aircrafts, confirming what Olli Rehn stated, that "Turkey's integration into the European Union is a strategic issue"[84].

The immigration problem and the issue concerning FRONTEX's missions in the Aegean area has been a chronic issue. Resolving these matters have not been realised due to the fact that the disputes between Greece and Turkey still prevail. Athens has noted that each time immigration policies are the centre of talks within Europe, Ankara is pushed to take responsibility for the current situation and pressurised to adopt measures to solve it. Even so, it is imperative to state that Turkey has not yet done so. The latter was and still is against FRONTEX, for the reason that when aerial patrols started over the Aegean, images of the Turkish

---

[84] European Union, "Why Turkey and the EU need each other: co-operating on energy and other strategic issues", 5 June 2007, http://europa.eu/rapid/pressReleasesAction.do?reference=SPEECH/07/362&format=HTML&aged=0&language=EN&guilanguage=en, accessed 14/07/2010, 17.24

coastguard accompanying the vessels which smuggled immigrants were revealed and sent to the EU, emphasizing how Turkey is not contributing to the resolution of this regional and European on-going problem, but facilitating it, creating thus problems for the EU. It is apparent that by sending immigrants to Greece, which does burden the Greek economy, seen by many as a way of intensifying the Aegean dispute, Turkey is evidently also harming and causing many difficulties for the EU, a union which it wishes to join.

FRONTEX is not the only body intercepted by Turkey. In the second half of 2009 thirty two cases were recorded where planes from other European countries (from Slovakia, Luxembourg, Latvia, Sweden, Finland and Estonia), which were involved in illegal immigration patrols, were harassed by Turkey, whilst flying over the Greek island Farmakonisi. Turkish authorities claimed that they were flying in Turkish airspace. This case depicts the reality that not only the EU, as a whole, has negative consequences due to the on-going and chronic problem of the Aegean dispute, but also various states within the Union are inevitably brought in to it.

If these actions continue and the EU and Turkey do not come to an agreement on the matter of immigration, the harassment towards FRONTEX and other patrol aircrafts of the European Union will be intensified. The influx of thousands of illegal immigrants into the EU every year will inescapably produce many problems for the Union's cohesion.

This is an issue directly affecting Europe which is also a straight consequence of the Aegean dispute between Greece and Turkey. Consequently, Europe should attempt and mediate in solving this continual quandary. The results will be evident not only between the two neighbouring states, but also within the EU.

B. Regional Stability. The Border Issue

The second consequence of the Aegean dispute relates to the first one, since it refers to regional stability. Through the FRONTEX and the immigration issue, it is evident that the regional stability between the two neighbouring allied states is compromised. This, inevitably, also affects the European Union.

The major reason for this second consequence is that Europe has not to this day, as Greece wished and tried to implement, delimitated the borders of Europe (and therefore Greece's) in the Aegean. However, it is evident that if Europe did resolve the border issue then it would be considered an official mediator and would contribute towards the gradual solution of the Aegean dispute. Then again it could also have a negative effect, which Europe fears, i.e. Turkey would dislike any action taken by a third party in respect to any Aegean issue with Greece. Also any solution of this problem would violate Turkey's interests in the region. "Consequently, the dominant understanding of the Greek – Turkish border is as a physical line of territorial demarcation"[85], which subsequently means that Europe guards non-demarcated areas, producing problems for both Aegean states. However, it is apparent that even delimitation of the borders in the Aegean might not solve the problem between Greece (and the EU) and Turkey on the matter of regional security and illegal immigration. "Security concerns of Greece remain unaltered"[86] even though the EU and Greece have supported and made efforts for Turkey to become a member of the Union. Europe acknowledges Turkey as an important, strategic state, with an upcoming economy, making the latter state more important than Greece. This is also verified, since Turkey is part of the G20[87], playing an important role within the

[85] Rumelili, Bahar, 'The European Union's Impact on the Greek-Turkish Conflict', *The European Union and Border Conflicts: The Impact of Integration and Association (EU BorderConf),* No.6 (January 2004), p. 23

[86] Patsiaouras, Konstantinos, *Democratic Peace theory and Greek-Turkish Relations in the Context of the European Union,* (California, Thesis Naval Postgraduate School, 2009), p. 107

[87] "The Group of Twenty (G20) is the premier forum for its members' international economic cooperation and decision-making. Its membership comprises 19 countries plus the European Union. G20 leaders meet annually, additionally during the year Finance Ministers and Central Bank Governors meet regularly to discuss ways to strengthen the global economy, reform international financial institutions, improve financial regulation, and discuss the key economic reforms that are needed in each of the member countries. Underpinning these meetings is a year-long program of

global economic scene. Any future resolution would be positive for the EU but the Union seems unwilling to pressurise Turkey, since it understands that it is a primary regional power, which the West could use through relations and eventual union with the EU for its own benefit.

The border issue, which also entails regional stability, especially for both Greece and Turkey, is an issue which, if solved, would facilitate Europe battle illegal immigration, but also give it security in its Southern region. This border issue relates to the delimitation of the territorial seas and continental shelf, explained in the analysis of the Aegean dispute, making it economically expensive not only for Greece but also for Europe, since its 'unofficial' (at the moment) borders in the Aegean have to be safeguarded, in order to maintain the status quo, by preventing the arrival of new and countless immigrants on European soil.

The border matter and safety is relevant to the immigration issue explained above. However, this is a distinctive problem all together both for Greece, which is immediately affected by this current situation, and also the European Union, which has a problematic region within its geographical and political fold. This border issue has also deemed the relations with Turkey (undoubtedly important state

meetings among senior officials and of working groups coordinating policy on specific issues". G20, http://www.g20.org/about_G20, accessed 31/01/2014, 14.51

for Europe and the West from an economic, political and strategic point of view) problematic, by postponing its eventual integration into the Union.

C. The EU's Role in this Dispute Between Greece and Turkey

The EU has always observed the relations between Greece and Turkey in a critical manner. It had distanced itself equally from both states on the disputed issues. However, this eventually changed when Greece became a full member state in 1981. Now the dispute was between "an 'insider' and 'outsider'"[88], which had consequences for the Union. Ankara saw this as a future problem, especially when becoming a member of the EU became Turkey's political goal, which was the case until 1999, when Greece changed its stance towards Turkey's accession. Conversely, Greece has in a sense "Europeanized"[89] the Greek-Turkish dispute, to Turkey's disappointment.

The EU, although not taking sides or helping dynamically to resolving the Aegean dispute as a

[88] Onys, Zyya, "Greek-Turkish Relations and the European Union: A Critical Perspective", Koc University Istanbul, http://home.ku.edu.tr/~zonis/turkey-greece.PDF, accessed 19/05/2010, 18.47, p. 9

[89] Akiman, Nazmi, 'Turkish - Greek Relations: From Uneasy Coexistence to Better Relations? A Retired Ambassador Takes Stock', *Mediterranean Quarterly* 13.3, (2002), p. 30

third party, has kept neutral but has also maintained interest in this dispute. A solution of the issues between the two countries would facilitate not only Greece and Turkey but also the EU. For many decades the European Union has kept far away from the Greek-Turkish disputes, "leaving Western interventions either to NATO, the United States, or isolated diplomatic activities by some of its member countries"[90], as was the case during the Imia crisis in 1996, when neither the European Union nor any member state intervened or assisted in defusing the situation, a responsibility that the US president initiated. As Mr Holbrooke explains, who was in direct contact with both governments (when the Imia crisis occurred) trying to defuse the situation, "while President Clinton was on the phone with Athens and Ankara, the Europeans were literally sleeping through the night. You have to wonder why Europe does not seem capable of taking decisive action in its own theatre"[91]. This is the perfect case where not only Europe seems unwilling and incapable of acting when a crisis occurs but it also emphasizes, in a way, its disunity especially when a member state is on the brink of war. It also confirms

[90] Constas, Dimitris, *The Greek-Turkish Conflict in the 1990s-Domestic and External Influences,* (London, Macmillan, 1991), p. 162

[91] Pratt ,Martin and Clive, Schofield, 'The Imia / Kardak Rocks Dispute in the Aegean Sea', *IBRU Boundary and Security Bulletin Spring,* (1996), p. 68

that "the EU seems to be unable to determine its own security and defence arrangements"[92], when an external threat to a member state of the Union is provoked, highlighting Europe's reliance towards the USA and NATO.

On the other hand, Europe has clearly and rightfully stated that "a candidate country cannot hope to become a full-member of the Union as long as bilateral conflicts with an existing member persist"[93]. This inevitably will put pressure on Ankara to solve any difference it has not only with Greece over the Aegean dispute but also with a second member state, namely Cyprus over the issue of illegal invasion and occupation of nearly half of the island since 1974. Additionally, Turkey seems to have found an obstacle in its accession within the European Union. Despite wishing to become an EU member, Turkey still does not recognise the Republic of Cyprus, i.e. a member state of the European Union. This is a paradox. The recognition of Cyprus by Turkey is one of the requirements for its accession within the Union; without this Cyprus can veto

[92] Galiatsatos, Gerasimos, *Is There a Common Foreign, Security and Defence Policy of the European Union? EU Policy Towards Cyprus and the Western Balkans Compared,* (London, Dissertation University of North London, School of Language and European Studies, 2002), p. 13

[93] Onys, Zyya, "Greek-Turkish Relations and the European Union: A Critical Perspective", Koc University Istanbul, http://home.ku.edu.tr/~zonis/turkey-greece.PDF, accessed 19/05/2010, 18.47, p. 14

Turkey's accession at any time, showing that Greece is not currently Turkey's obstacle. Also, due to the change within the policy of Greece on this matter, other member-states have shown their concerns in regards to Turkey's full accession within the EU, including France and Germany, whilst other states, i.e. United Kingdom, show their support to this reality. Thus, due to the above requirement, it is evident that Europe has made a small move to finally resolve the disputed issues between Greece and Turkey; nevertheless, it still has not made clear its views on the matter.

The conduct and the positions held by the European Union concerning the Aegean dispute between Greece and Turkey stress Europe's incapability to solve a member state's disputed issues with a neighbouring state. The EU maintains a neutral stance, whilst accepting the current status quo of relations between the two Aegean countries, "since they have interests in both countries"[94]. This was also the case with Cyprus, where the current status quo was not acceptable for the EU, which was one reason to keep the island out of the Union. However, this altered, due to Greece's possibility of vetoing further expansion towards Eastern Europe and halting any new accessions for the Union; therefore, Cyprus was accepted in the EU in 2004,

[94] Platias, Athanasios G., 'Greek Deterrence Strategy', in Chircop, Aldo, Andre, Gerolymatos and John, Iatrides (Eds), *The Aegean Sea after the Cold War- Security and Law of the Sea Issues,* (London. Macmillan Press LTD., 2000), p. 71

due to Greece's pressure and vetoing power. In spite of this, other reasons, mainly economic, political and strategic, will be the eventual reason for Turkey's acceptance to the Union in the future. The disputed matters with Turkey for the two member states (Greece and Cyprus) will probably be disregarded, resulting in the alteration of European practices (as was the case with NATO).

The EU, in the near future, will have to play a key role in mediating in the Aegean dispute between Greece and Turkey, more than it has done these past decades, especially since 1981 when Greece became a full member of the Union. Also it must act in a greater manner than the USA and NATO, since this dispute exists in its geographical region, making it inevitably a European problem. However, the eventual consequence of this dispute between a member state and a non-member state for the EU is that it shows clearly its neutral stance towards individual problems that occur between member states. Even so, if this dispute is not settled before Turkey becomes a member state of the EU then "the conflict between the two countries would be transferred to a Community context"[95], which would have unforeseen affects for the Union. Nonetheless, this consequence of maintaining a neutral stance and also being interested in the dispute, is a policy kept

---

[95] Constas, Dimitris, *The Greek-Turkish Conflict in the 1990s-Domestic and External Influences,* (London, Macmillan, 1991), p. 216

by the EU, since it ignores disputes "concerning national matters"[96], as was for example the case with the UK and Spain over Gibraltar. Thus, this last consequence exists not only specifically concerning the dispute in the Aegean, but is a rather general passive policy maintained by the Union and its members, emphasising therefore its disunity on a number of matters.

Conclusion

Concluding this chapter, it is apparent that not only Greece and Turkey have negative consequences due the Aegean dispute but also the European Union feels the effects of the bilateral disagreements between a member state and a non-member state. The three consequences for the Union analysed above (the immigration and FRONTEX matter, the regional stability which relates to the border issue and the EU's role in this dispute) emphasize how important this dispute is for the European Union, not only politically but also economically and strategically. The continuation of this chronic dispute will only keep these problems alive for Greece, Turkey and the EU, and it will probably worsen their relations.

[96] Constas, Dimitris, *The Greek-Turkish Conflict in the 1990s-Domestic and External Influences,* (London, Macmillan, 1991), p. 223

A future involvement by the EU as a third party, due to its international status, will be a positive move both for the Union's cohesion but also for both states individually. However, if the European Union continues accepting the consequences from the Aegean dispute it will just maintain the problems which occur from it. Also its cohesion will probably be compromised, as it becomes evident in the case of Greece, which is a member state of the EU but yet feels "betrayed by its European allies and partners and to a significant degree isolated"[97].

The EU has to act in a decisive way and preserve its cohesion and integrity, and not allow the USA and NATO to play a protagonistic role in solving European problems and disputes. This will strengthen the Union. Although it is understandable how other EU member states might seem "reluctant to get dragged into any war"[98] or disputed matter between Greece and Turkey, any action taken by the EU will be a positive move towards its security.

---

[97] Athanasopoulos, Haralambos, *Greece, Turkey and the Aegean Sea*, (Jefferson, North Carolina, McFarland and Company Inc. Publishers, 2000), p. 94

[98] Barchard David, "Turkey and the European Union", Centre for European Reform, http://www.cer.org.uk/pdf/p093_turkey.pdf , accessed 10/08/2010, 20.02, p. 8

GENERAL CONCLUSION

The Aegean dispute between Greece and Turkey is a chronic disagreement between two NATO allies. It is an ongoing quarrel over many issues concerning the status quo in the Aegean. The continental shelf issue is the only matter which Greece recognizes as a bilateral problem between the two countries. Turkey on the other hand introduces other disputed topics which complicate the whole Aegean issue, namely the territorial seas, airspace, demilitarization and militarization of the Eastern Aegean Islands and the disputed islands, islets, rocks, which leads to the recent grey zones theory that Ankara introduced in 1996. All of these matters prevail due to the "mutual suspicion and mistrust"[99], making it impossible for a resolution of this dispute to be found in the foreseeable future.

All of these individual problems that are part of the greater Aegean problem explain why Greece and Turkey are in dispute over the sea that unites but also separates them. Economic, political, strategic reasons, especially the fact that the Aegean is "a very

---

[99] Aydin, Mustafa, "Contemporary Turkish-Greek Relations: Constraints and Opportunities", in Aydin, Mustafa and Kostas, Ifantis (eds), *Turkish-Greek Relations, The Security Dilemma in the Aegean*, (London and New York, Routledge, 2004), p. 45

important shipping route"[100], are all the fundamental grounds for the prevalence of this dispute between the two allied states. Inevitably, whoever controls the Aegean Sea will be able to increase their profitability, commerce and geo-political power, becoming the most significant, strategically, power within that region of the world. However, as it has been suggested, "the mutual politico-economic gains of an ensuing resolution vastly outweigh the individual politico-military desires of deadlock"[101]. This is why a third party, such as NATO and the EU or maybe the UN and the USA, would be the key to a quick and impartial resolution of this chronic quarrel.

If we are to maintain a neo-realist view of this bilateral matter then it is evident that "international politics is not a constant state of war, but it is a state of relentless security competition"[102], which makes war always a possibility. This can only be solved if both Greece and Turkey try to find a solution to

---

[100] Dokos, P. Thanos, "Tension-Reduction and Confidence-Building in the Aegean", in Aydin, Mustafa and Kostas, Ifantis (eds), *Turkish-Greek Relations, The Security Dilemma in the Aegean*, (London and New York, Routledge, 2004), p.125

[101] Mann, Stephen, *The Greek-Turkish Dispute in the Aegean Sea: Its Ramification for NATO and the Prospects for Resolution*, (Monterey California, Naval Postgraduate School, 2001), p. 78

[102] Aydin, Mustafa and Kostas, Ifantis (eds), *Turkish-Greek Relations, The Security Dilemma in the Aegean*, (London and New York, Routledge, 2004), p. 291

these problems, which do not only affect them, but also affect other international bodies, such as NATO and the EU. It is apparent that the solution of many, if not all, the Greek-Turkish disputes will pave the way for a brighter future for both the Aegean countries, short term and long term, clearing the decks for them both to play a strategically important role within the region.

Greece and Turkey are not the only ones who observe and inevitably receive the negative consequences of this dispute. As was evident in the previous chapters, NATO and the European Union have also been affected by this dispute. The consequences analysed in the previous chapters for both international bodies give emphasis to the fact that this quarrel has troubled the international community and a rapid and an effective resolution of these issues between the two states would certainly be a welcoming and positive achievement for both NATO and the EU. However, it is obvious that this dispute between Greece and Turkey is not the biggest issue concerning NATO at this very moment. The war in Iraq and Afghanistan are much more important and crucial for the alliance, whilst issues in the Middle East and North Africa have been troubling the international community for years. On the other hand, it is not dismissed as a small problem within the alliance, especially if we take into consideration the fact that the South-Eastern flank is vulnerable due to this dispute, since the military exercises over the Aegean are always accompanied by challenges, which are produced by either Greece or Turkey. The reality of not being able to take vital decisions, which need a unanimous vote, due to their dispute, have paralysed NATO and prevented it from becoming a better international strategic institution, as was evident in chapter 2. Also the fact that its current stance towards this dispute between two of its members has been neutrality underlines a weakness and disunity within the alliance.

The European Union, on the other hand, furthermore experiences negative consequences of the Aegean dispute, on a similar level as NATO. The pan-European problem of illegal immigration, which burdens all the economies in the EU, is the prominent consequence of the dispute between Greece and Turkey. Illegal immigration is directly related to another problem, that of FRONTEX being not always free and able in policing the Aegean. Turkey has constantly produced inconveniences on this matter due to the Aegean dispute. Also regional stability and the border issue are relevant not only to the two neighbouring states but also to the cohesion and security of the Union. This makes the South-Eastern part of Europe a challenging region. The EU's role in this dispute between a member state and a non-member state is crucial, since it will show to the world, but also to its own members, that cohesion and security are important for the continuation and the progression of the Union. Additionally, this disputed matter has also revealed the EU's weaknesses in solving or at least assisting in creating a better inter-state climate.

Although "Greek and Turkish leaders do not anticipate a resolution of the Aegean disputes in the foreseeable future"[103], it is apparent that a solution has to be found, whether it is on a bilateral level, which Turkey supports, or by intervention of a third party in the international field, as Greece proposes. A first step to achieving peace and resolving the Aegean dispute between the two allied states is for both to "address and understand each other's justifiable threats, fears and national security concerns"[104]. A final resolution of the Aegean dispute will have positive and profitable results for both Greece, by strengthening its position in the EU and NATO, and Turkey, by strengthening its position in NATO and by eventually becoming a full member of the European Union.

---

[103] Bahcheli, Tozun, "The Potential for Conflicts in Greek-Turkish Relations, Austrian Armed Forces", http://www.blmv.gv.at/pdf_pool/publikationer/03_jb00_24.pdf, accessed 27/05/2010, 14.58

[104] Athanasopoulos, Haralambos, *Greece, Turkey and the Aegean Sea*, (Jefferson, North Carolina, McFarland and Company Inc. Publishers, 2000), p. 55

BIBLIOGRAPHY

Acer, Yucel, *The Aegean Maritime Disputes and International Law*, (Burlington, Ashgate Publishing Limited, 2003)

Akiman, Nazmi, 'Turkish - Greek Relations: From Uneasy Coexistence to Better Relations? A Retired Ambassador Takes Stock', *Mediterranean Quarterly* 13.3, (2002)

Anastasakis, Othon, "Greek - Turkish relations: Old problems, new challenges", Barcelona Centre for International Affairs , www.cidob.org, accessed 25/05/2010, 17.35

Antoniou, Dora, "Epikindines Tourkikes prokliseis sto Aigaio, Apo tis parenohlisis aeroskafon tis FRONTEX meso rantar, stis epihirisis anaxaitisis elikopterou tou Limenikou apo F-16", *KATHIMERINI,* http://news.kathimerini.gr/4dcgi/_w_articles_2_28/03/2010_395803, accessed 28/03/10, 14.56

Athanasopoulos, Haralambos, *Greece, Turkey and the Aegean Sea*, (Jefferson, North Carolina, McFarland and Company Inc. Publishers, 2000)

Aydin, Mustafa, "Cacophony in the Aegean; Contemporary Turkish Greek Relations", University of Ankara, http://www.politics.ankara.edu.tr/eski/dosyalar/MMTY/27/6_mustafa_aydin.pdf, accessed 28/05/2010, 17.46

Aydin, Mustafa, "Contemporary Turkish-Greek Relations: Constraints and Opportunities", in Aydin, Mustafa and Kostas, Ifantis (eds), *Turkish-Greek Relations, The Security Dilemma in the Aegean*, (London and New York, Routledge, 2004)

Aydin, Mustafa and Kostas, Ifantis (eds), *Turkish-Greek Relations, The Security Dilemma in the Aegean*, (London and New York, Routledge, 2004)

Bahcheli, Tozun, "The Potential for Conflicts in Greek-Turkish Relations", Austrian Armed Forces, http://www.blmv.gv.at/pdf_pool/publikationer/03_jb00_24.pdf, accessed 27/05/2010, 14.58

Bahcheli, Tozun, Theodore A., Couloumbis and Patricia, Carley, *Greek-Turkish Relations and U.S. Foreign Policy - Cyprus, the Aegean and Regional Stability*, (Washington, United States Institute of Peace, 1997)

Barchard David, "Turkey and the European Union", Centre for European Reform, http://www.cer.org.uk/pdf/p093_turkey.pdf , accessed 10/08/2010, 20.02

Bolukbasi, Deniz, *Turkey and Greece The Aegean Disputes- A Unique Case in International Law*, (London, Cavendish Publishing Limited, 2004)

Chircop, Aldo, Andre, Gerolymatos and John, Iatrides (Eds), T*he Aegean Sea after the Cold War- Security and Law of the Sea Issues,* (London. Macmillan Press LTD., 2000)

Constas, Dimitris, *The Greek-Turkish Conflict in the 1990s- Domestic and External Influences,* (London, Macmillan, 1991)

Davutoglu, Ahmet, *To Stratigiko Vathos, I Diethni Thesis tis Tourkias*, (Athens, Poiotita, 2010)

Dokos, P. Thanos, "Tension-Reduction and Confidence-Building in the Aegean", in Aydin, Mustafa and Kostas, Ifantis (eds), *Turkish-Greek Relations, The Security Dilemma in the Aegean*, (London and New York, Routledge, 2004)

Doyle, Michael W., 'Kant, Liberal Legacies and Foreign Affairs', *Philosophy and Public Affairs,* Vol. 12, No. 3, (Summer, 1983), p. 205-235

Ellis, Athanasios and Michalis, Ignatiou, *IMIA-TA APORRITA TILEGRAFIMATA TON AMERIKANON*, (Athens, Livani, 2009)

Frontex, http://frontex.europa.eu/about-frontex/mission-and-tasks, accessed 31/01/14, 15.28

G20, http://www.g20.org/about_G20, accessed 31/01/2014, 14.51

Galiatsatos, Gerasimos, *Is There a Common Foreign, Security and Defence Policy of the European Union? EU Policy Towards Cyprus and the Western Balkans Compared,* (London, Dissertation University of North London, School of Language and European Studies, 2002)

Heraclidis, Alexis, "The Greek-Turkish Conflict: Towards Resolution and Reconciliation", in Aydin, Mustafa and Kostas, Ifantis (eds), *Turkish-Greek Relations, The Security Dilemma in the Aegean*, (London and New York, Routledge, 2004)

Iatrides, O. John, "NATO and Aegean Disputes: The Cold War and After", in Chircop, Aldo, Andre, Gerolymatos and John, Iatrides (Eds), *The Aegean Sea after the Cold War-Security and Law of the Sea Issues*, (London. Macmillan Press LTD., 2000)

Jackson, Alex, "Natural, Gas Europe" http://www.naturalgaseurope.com/greece-and-turkey-spar-over-aegean-energy, accessed 18-09.2013, 20.29

Kalaitzaki, Theodora, "US Mediation in Greek-Turkish Disputes since 1954", *Mediterranean Quarterly* 16.2 (2005), . 106-124

Karaosmanoglu, Ali, "Turkey and the Southern Flank: Domestic and External Contexts", in Chipman, John (Eds.), *NATO's Southern Allies: Internal and External Challenges,* (London and New York, Routledge, 1988)

Mann, Stephen, *The Greek-Turkish Dispute in the Aegean Sea: Its Ramification for NATO and the Prospects for Resolution*, (Monterey California, Naval Postgraduate School, 2001)

Migdalovitz, Carol, "Turkey: Issues for U.S. Policy", Congressional Research Service, http://www.fas.org/man/crs/RL31429.pdf, accessed 20/05/2010, 15.26

Morris, Chris, "Analysis: Hostility Returns to Aegean", BBC news, http://news.bbc.co.uk/1/hi/world/europe/984135.stm, accessed 04/08/2010, 20.14

Moustakis, Fotios, *The Greek - Turkish Relationship and NATO,* (London, Frank Cass Publishers, 2003)

Onys, Zyya, "Greek-Turkish Relations and the European Union: A Critical Perspective", Koc University Istanbul, http://home.ku.edu.tr/~zonis/turkey-greece.PDF, accessed 19/05/2010, 18.47

Patsiaouras, Konstantinos, *Democratic Peace theory and Greek-Turkish Relations in the Context of the European Union*, (California, Thesis Naval Postgraduate School, 2009)

Paulenoff, Alex, "The Aegean sea Continental Shelf Dispute: Greek and Turkish Interpretations", Perspectives on Global Issues, http://www.perspectivesonglobalissues.com/archives/spring-2009/continental, accessed 23/05/2010, 19.19

Platias, Athanasios G., 'Greek Deterrence Strategy', in Chircop, Aldo, Andre, Gerolymatos and John, Iatrides (Eds), *The Aegean Sea after the Cold War- Security and Law of the Sea Issues,* (London. Macmillan Press LTD., 2000)

Pratt ,Martin and Clive, Schofield, 'The Imia / Kardak Rocks Dispute in the Aegean Sea', *IBRU Boundary and Security Bulletin Spring,* (1996)

Prescott, J.R.V., *The Maritime Political Boundaries of the World,* (London and New York, Methuen & Co, 1985)

Rumelili, Bahar, 'The European Union's Impact on the Greek-Turkish Conflict', *The European Union and Border Conflicts: The Impact of Integration and Association (EU BorderConf),* No.6 (January 2004)

Strati, Athanasia, "Greece and the Law of the Sea: A Greek Perspective", in Chircop Aldo, Andre Gerolymatos and John Iatrides (eds), *The Aegean Sea After the Cold War* (London, Macmillan Press LTD., 2000)

Syrigos, M. Angelos, *The Status of the Aegean Sea According to International Law*, (Athens, Bruxelles, Sakkoulas/Bruylant, 1997)

Syrmos, Theodoros, *Territorial Waters of Greece and Air/Maritime Navigation*, (Alabama, Maxwell Air Force Base, 2002)

Tsakonas, Panayiotis, Weatherhead Center for International Affairs, "Greek-Turkish Relations Towards the 21st Century- A View from Athens", http://www.wcfia.harvard.edu/fellows/papers/1998-99/tsakonas.pdf , accessed 30/05/2010, 16.10

Turgut, Pelin, 'Turning to the East', Time Magazine, (July 5, 2010), p. 26-29

European Union, "Why Turkey and the EU need each other:co-operating on energy and other strategic issues", 5 June 2007, http://europa.eu/rapid/pressReleasesAction.do?reference=SPEECH/07/362&format=HTML&aged=0&language=EN&guilanguage=en, accesed 14/07/2010, 17.24

Global Security, "Destined Glory", http://www.globalsecurity.org/military/ops/destined-glory.htm, accessed 13/07/2010, 16.01

Hellenic National Defence General Staff, http://www.geetha.mil.gr/index.asp?a_id=2781, accessed 26-07-2010, 11.56

Hurriyet Daily News, "Greece Says Turkish F-16s Harass EU Immigration Patrols", http://www.hurriyetdailynews.com/n.ph?n=greece-says-turkish-f-16s-harass-eu-immigration-patrols-2010-03-19, accessed 19/07/2010, 16.48

International Boundary Study, "Greece-Turkey Boundary", http://www.law.fsu.edu/library/collection/LimitsinSeas/IBS041.pdf, accessed 28-07-2010, 19.02

Kathimerini newspaper, "Diavima apo to YPEKS gia 'Tsesme' kai 'Piri Reis', http://www.kathimerini.gr/4dcgi/_w_articles_kathremote_1_20/07/2010_347532, accessed 20-07-2010, 13.47

Kathimerini, "Turkish jets chased off after violating Greek air space", http://www.ekathimerini.com/4dcgi/_w_articles_wsite1_1_20/08/2013_514765, accessed 18/09/2013, 20.12

Ministry of Foreign Affairs, Greece, http://www.ypex.gov.gr/www.mfa.gr/en-US/Policy/Geographic+Regions/South-Eastern+Europe/Turkey/Relationships+-+Potentials/, accessed 26-07-2010, 12.02

Ministry of Foreign Affairs-Turkey, http://www.mfa.gov.tr/background-note-on-aegean-dispute.en.mfa, accessed 31-07-2010, 19.43

NATO, "Lisbon Summit Declaration", http://www.nato.int/cps/en/natolive/official_texts_68828.htm?mode=pressrelease, accessed 01/10/2013, 09.30

NATO, "Statement by the Secretary General of NATO, Dr. Javier Solana, on Confidence Building Measures Between Greece and Turkey", http://www.nato.int/cps/en/natolive/news_25854.htm?selectedLocale=en, accessed 04/08/2010, 21.36

Smith, Helena and Ian Traynor, "Mid-air fighter plane collision risks new Greek-Turkish crisis", The Guardian, http://www.theguardian.com/world/2006/may/24/turkey.topstories3, accessed 02-10-2013, 13.10

The North Atlantic Treaty, http://www.nato.int/cps/en/natolive/official_texts_17120.htm, accessed n05/08/2010, 16.58

Time Magazine, "THE MEDITERRANEAN: The West's Ragged Edge", 12 June 1978, http://www.time.com/time/magazine/article/0,9171.948171,00.html, accessed 25-05-2010, 18.46

Treaty of Lausanne, http://wwi.lib.byu.edu/index.php/Treaty_of_Lausanne, accessed 02/08/2010, 16.28

U.S Arms Clients Profile-Greece, http://fas.org/asmp/profiles/greece.htm, accessed 14/07/2010, 14.53

THE AUTHOR

Dimitris Salapatas has studied Theology at the National and Kapodistrian University of Athens, School of Theology, as well as Byzantine Music at the National College of Music (Athens). He then studied International Relations on a Postgraduate level at Queen Mary, University of London, where he wrote his thesis, examining the question, "Why are Greece and Turkey in dispute over the Aegean Sea and what consequences does it have for both NATO and the EU?" He later did two internships at the: a. Press

Office of the Greek Embassy in London and b. The Office of MP Barry Gardiner (at both the local office - Brent - and in Parliament). He is currently undertaking a research on the "Fellowship of St. Alban and St. Sergius and its contribution towards Anglican - Orthodox Relations", at the University of Winchester. Dimitris has published numerous articles in both English and Greek on various topics and has given talks on various themes, including Anglican-Orthodox Relations (London) and Nationalism vs. Ecumenism (Oxford). The author is currently a Committee Member of the Anglican and Eastern Churches Association (A.E.C.A.), representing the Greek Orthodox Archdiocese of Thyateira and Great Britain, Committee Member of the Educational Sub-Committee of the Pan-Orthodox Episcopal Assembly for the British Isles and Ireland, Committee Member of the Peloponnesian Association of Great Britain and member of the Fellowship of St. Alban and St. Sergius. He is currently a chanter (Lambadarios) at the Greek Orthodox Cathedral Church of All Saints, Camden Town, London.

Printed in Dunstable, United Kingdom

91342776R00067